Reliability Analysis of Dynamic Systems: Efficient Probabilistic Methods and Aerospace Applications

Elsevier and Shanghai Jiao Tong University Press Aerospace Series

T0259092

Reliability Analysis of Dynamic Systems: Efficient Probabilistic Methods and Aerospace Applications

Elsevier and Shanghai Jiao Tong University Press Aerospace Series

Bin Wu

AMSTERDAM • BOSTON • HEIDELBERG • LONDON
NEW YORK • OXFORD • PARIS • SAN DIEGO
SAN FRANCISCO • SINGAPORE • SYDNEY • TOKYO

Academic Press is an imprint of Elsevier

Academic Press is an imprint of Elsevier
225 Wyman Street, Waltham, MA 02451, USA

British Library Cataloguing-in-Publication Data
A catalogue record for this book is available from the British Library

Library of Congress Cataloging-in-Publication Data
A catalog record for this book is available from the Library of Congress

ISBN: 978-0-12-407711-9

For information on all Academic Press publications
visit our website at elsevierdirect.com

Typeset by MPS Limited, Chennai, India
www.adi-mps.com

Printed and bound in the US

13 14 15 16 17   10 9 8 7 6 5 4 3 2 1

Working together
to grow libraries in
developing countries

www.elsevier.com • www.bookaid.org

Contents

To my family

Deterministic analysis approaches/tools have dominated the whole aerospace industry for many years. It has been widely accepted, however, that the relevant non-deterministic analysis methods, either probabilistic or possiblistic, will be eventually adopted to some extent in this area. This process has been very slow, partly due to the conservative nature of the industry and partly due to some difficulties in applying these methods, which are now being addressed by both academia and industry.

Within the last decade in the engineering field, possibilistic approaches have been widely studied and applied to the reliability analysis of dynamic systems. During this period, there has been a lack of research interest in delivering efficient probabilistic methods. This book presents a novel technique that applies probabilistic methods to reliability analysis of engineering systems under harmonic loads in the low-frequency range. The aim was to overcome certain problems of applying probabilistic methods. The problems that need to be overcome were the nonlinearity of the failure surface, the intensive computational cost, and the complexity of the dynamic system.

A perturbation analysis algorithm was developed based on a modal approximation model. Since the resonance cases are of most concern, the optimized model simplifies the complexity of the dynamic systems by only concentrating on the resonance dominating terms in the response element (expressed in terms of modal coordinates). This optimization and later newly defined parameters transform the original failure surface into an approximate but smooth and linear one. Finally, the statistical information of the new parameters can be derived from that of the original variables by solving only once the eigen problem on the mean values of the original variables. An efficient reliability method, such as FORM, can then be applied.

However, for a given 2D frame structure, the FORM method failed to accurately predict the probability of failure. The Monte Carlo simulation method was later adopted to replace the FORM method. The Monte Carlo simulations were only performed for the new random parameters that were obtained through one execution of an eigen solver. Thus the overall efficiency of this combined approach, i.e. perturbation approach plus Monte Carlo simulation method, is high. Both accuracy and efficiency were achieved when this combined approach was applied to the 2D structure, as well as to a complex 3D helicopter model. Finally the response surface method was employed to derive the statistical information of the stiffness matrix from that of the original property random variables.

Low modal overlap factor, responses near resonance, low statistical overlap and small changes in eigenvalues and Gaussian distribution of the original variables are the conditions required for this approach to work.

Acknowledgments

My sincere thanks are firstly due to Professor Robin S. Langley, my supervisor during the years in Cambridge, for providing academic ideas, patient guidance and valuable support. The advice and help that I received from Sondipon Adhikari, Srikantha Phani, Andrew Grime, Rolf Lande, Brian Jujnovich, Simon Rutherford and other members in the Dynamics and Vibration Research Group will not be forgotten.

The support and information given freely and generously by researchers in the engineering domain outside Cambridge are acknowledged with much gratitude, in particular, Dr Qin Feng and Dr Jim Margetson, whose names should be mentioned.

My due thanks go to my colleagues at the Commercial Aircraft Corporation of China, Ltd (COMAC). Frequent discussion with Dr Qian Guo, Shanghai Aircraft Design and Research Institute of COMAC, was technically very useful. Mr Xiaojun Xue and Mr Peng Wang deserve my special thanks for the information and expertise they provide on engineering reliability, aviation safety and airworthiness. I would like to express my sincere thanks to Mr Qingwei Zhang, former Board Chairman of COMAC, Mr Zhuanglong Jin, current Board Chairman of COMAC, Mr Hua Yan, Director of HR department of COMAC, and Mr Fuguang Qin, Director of Beijing Research Centre of COMAC, for their help and support of my research work.

I am very grateful to the Engineering and Physical Science Research Council (UK), QinetiQ, Queens' College Cambridge and COMAC for funding my research. I express my sincere gratitude to Shanghai Jiao Tong University Press and Elsevier Limited for publishing this book.

I would like to thank my parents, my brother and sister-in-law, for their eternal love, constant support and encouragement that are of great value to me to overcome many challenges and difficulties in life. My special thanks go to my wife, Dr Jianxiang Cao, and my children, for their love and time. I am also grateful to my friends in Cambridge, London, Manchester, Beijing, Shanghai and Taibei for their advice and help that I received when needed.

Dr Bin Wu
COMAC, China
March 2013

ABBREVIATIONS

ACSR	Active control of structural response
AVS	Active vibration suppression
AVC	Active vibration control
BG	Bubnov–Galerkin
DOF	Degree of freedom
FE	Finite element(s)
FEA	Finite element analysis
FEM	Finite element method
FFEM	Fuzzy finite element method
FORM	First-order reliability method
FRF	Frequency response function
GOE	Gaussian orthogonal ensemble
HHC	Higher harmonic control
IBC	Individual blade control
jpdf	Joint probability density function
MC	Monte Carlo (simulation method)
MCS	Monte Carlo simulation (method)
pdf	Probability density function
PDE	Partial differential equation
RS	Response surface
RSM	Response surface method
SEA	Statistical energy analysis
SFE	Statistical finite element
SORM	Second-order reliability method
SRBM	Stochastic reduced basis method
TEF	Trailing edge flap

NOTATION AND SYMBOLS

M	Mass matrix
K	Stiffness matrix
A	Area
E	Modulus of elasticity (Young's modulus)
L	Length
β	Safety index
ρ	Property density
η	Loss damping factor
ω	Radian frequency/excitation frequency
f	Cyclic frequency (Hz)/excitation frequency

$[\Phi]$	Mass-normalized modal matrix
ϕ_j	jth column vector of mass-normalized modal matrix
ω_i	ith undamped natural frequency
$\{\psi_i\}$	ith mode shape
$P(\ )$	Probability
f_x	Pdf of random variable x
μ_x	Mean value of random variable x
σ_x	Standard deviation of random variable x
$E(x)$	Expected value of random variable x
$D(x)$	Variance of random variable x
$C_x(Cov_x)$	Covariance matrix of random variable x
C	Confidence level
α	Fuzzy confidence level
Φ	Standard normal distribution function

List of Figures

List of Tables

Chapter 1

Introduction

It is widely believed that, to investigate structural behaviour and reliability for the design of large engineering products such as aircraft or offshore structures, resorting to probabilistic methods is a necessity. However, this fact has only been recognized recently, or to be more accurate, it was only a few decades ago that it became possible to implement probabilistic reliability approaches in engineering practice. This chapter presents a brief history of structural reliability analysis, as well as modern analysis methods and the scope of current research work.

1.1 STRUCTURAL RELIABILITY ANALYSIS

Geometrical harmony of structures is a paramount condition for many ancient and medieval buildings to survive through hundreds of years. Examples are seen across different civilizations in different periods of time, for example Greek temples, Gothic cathedrals, to oriental palaces, and the Great Wall. There is no doubt that the ancient architects, in different nations, had an intuitive understanding that it is the *correct shape* of the structure that governs the stability in accommodating forces and stresses[1] satisfactorily [1]. Construction rules were derived from those intuitive understandings, and were recorded as early as Ezekiel around 600 BC, followed by many great inventions and corrections in the next 2000 years, for example *Vitruvius module* (c. 30 BC) [2] and *Mignot manual* (1400 AD) [1,3].

However, no design approaches for safety[2] were found in any of these early documents [1]. There was no prediction of any kind of failure; instead, the safety or stability was learnt, and finally satisfied, through modeling, trial and error. As a consequence, many structures did fail, particularly when dynamic complexity was involved. One of the examples was the story of the world famous seventeenth century Scandinavian ship – the Swedish Vasa – which capsized and sank on her maiden voyage in 1636 near Stockholm. 360 years later, it was discovered and salvaged. Scientific studies

---

1. It should also be noted that those structures, often built for holy, royal or military services, were made of high-quality materials at their time, which yield strength and durability.
2. Here, *safety* or *stability* is referred to as a pre-modern-era term sharing the same purpose but not the same content as the reliability concept, which was developed later to deal with probability.

Reliability Analysis of Dynamic Systems.
© 2013 Shanghai Jiao Tong University Press. Published by Elsevier Inc. All rights reserved.

that followed have revealed that lack of proper design and testing approaches for stability were the main reasons for this disaster. The ballast section at the bottom of the ship is shown in Figure 1.1. Studies found that the ballast section should have been much wider or deeper to load more stones/cargo to maintain stability.

The production of a reliable structure requires calculations in advance to predict the likelihood of structural failure. The understanding and evaluation of the strength of the materials and the relevant load effects, for either static or dynamic systems, is essential in such predictions. Around the time of the Vasa disaster, in-depth studies of structural behaviors and strength of materials were carried out by some world-leading physicists and mathematicians. One of them was Galileo, whose *Discorsi*, in the form of four dialogs, was published in 1636, in which (in the second dialog) observations and calculations of forces and stresses on some typical structure examples were reported [1]. Structural analysis developed dramatically in the following centuries. Many fundamental theories and scientific experimental systems were established. More accurate calculations became available, though some ideas might have been proposed and studied at an earlier time. An interesting example was the determination of the exact proportion of height and width of the cross-sectional area of a wooden rectangular beam structure that can yield the highest stiffness. A Chinese architect, Jie Li in the Song Dynasty, in his work *Construction Regulations* (1103 AD) [4], stated that the value should be 3:2, which was one of the closest estimates to the correct ratio in the history of structural analysis. This proportion was studied by many physicists thereafter, including Da Vinci and Galileo, until in 1807 Thomas

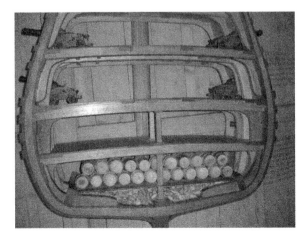

FIGURE 1.1   Cross-section of Vasa. Photo taken in the Vasa Museum, Stockholm, 3 October 2005. The bottom ballast weighs 120 tons in total, but it was still not enough to maintain the stability of the ship with two gundecks and the heavy artillery.

Young scientifically proved that the ratio is $\sqrt{2}$:1. The study of strength and load effects continues in modern times for many new materials and products.

The *Industrial Revolution* started in Great Britain in the nineteenth century. Many new technological inventions and engineering advancements emerged that revolutionized the lives of the whole of mankind thereafter. Reliability became of paramount importance. It is widely believed that it was during this period of time that the concept of a *safety factor* (or *load factor*) was introduced and implemented, though the history of all the early design/construction codes was lost [5]. The strength limit and the load limit were determined and augmented by safety/load factors in the design process to guarantee that there was no violation. This design approach is called the *deterministic approach*, as opposed to the later *probabilistic approach*, which deals with uncertainties probabilistically.

There is evidence, however, that the uncertainties of strength and loads were well recognized before the safety factor began to be implemented [5,6]. The deterministic approach does allow uncertainties but adopts a technique that separates the upper and lower boundaries of the random variables of interest. The effect of this technique is illustrated in Figures 1.2 and 1.3. The strength resistance (from an element of a structure) and the load effect are both considered uncertain, which can be described by a type of distribution density pattern, i.e. the *probability density function* (pdf). For example, in this case, a Gaussian distribution type is denoted as $f_R(r)$ and $f_S(s)$ for the strength resistance and load effect respectively, with mean values μ_R and μ_S. The overlapped area, the shaded area shown in the figure where the load effect exceeds the resistance, represents the *probability of failure*. Theoretically,

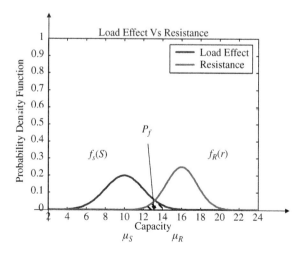

FIGURE 1.2 Probability distribution of the resistance and load effect of a member of a structure.

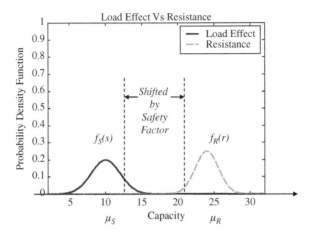

FIGURE 1.3 Safety factor separating sufficiently between the resistance and load effect.

there are a few techniques that can calculate the likelihood of occurrence of exceeding particular limits, such as integration methods or simulation methods. However, more than 100 years ago, it was difficult in practice to realize this, because firstly a working probabilistic approach was lacking, and secondly it was impossible to execute a large number of numerical simulations without the help of modern computational power. The deterministic approach was then the best solution.

In the theory of a deterministic approach, although the strength and loads are considered uncertain, it was accepted that an absolute upper limit to any load, and a lower limit to any strength, could be established. This is illustrated in Figure 1.3. A *safety factor* (or *load factor*)[3] is then applied to separate these limits "sufficiently" for every member of the structure. The uncertainties from the both sides were "ignored" or well accommodated by the shifting gap, which reserves enough strength for even the highest load effect. This would mean, in terms of the nature of the design (for an ultimate stress design or an ultimate strength design), either adopting higher strength (often implying choice of a better material) or prescribing a milder load effect.

For many years and for many engineering products, the deterministic solution worked so well that the whole engineering world once believed that the uncertainty problem had been scientifically fixed [5]. The reliability analysis of engineering structures was thereafter dominated by the deterministic approaches until the 1960s.

---

3. *Safety factors* (*load factors*) were selected, determined, and justified according to engineering, economic or sometimes even political considerations [7].

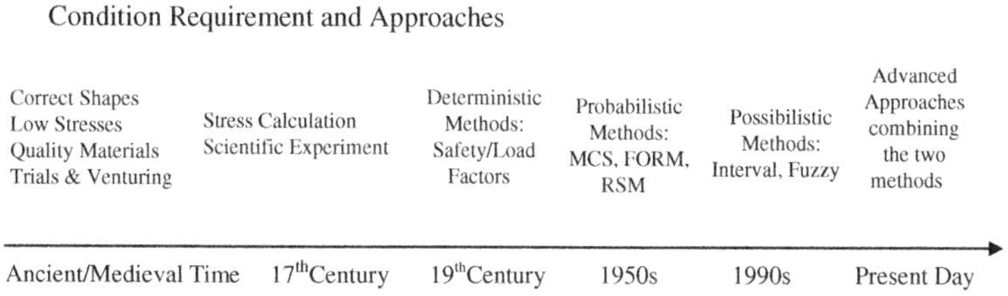

Condition Requirement and Approaches

Correct Shapes Low Stresses Quality Materials Trials & Venturing	Stress Calculation Scientific Experiment	Deterministic Methods: Safety/Load Factors	Probabilistic Methods: MCS, FORM, RSM	Possibilistic Methods: Interval, Fuzzy	Advanced Approaches combining the two methods
Ancient/Medieval Time	17thCentury	19thCentury	1950s	1990s	Present Day

FIGURE 1.4 Historical progress of structural reliability analysis approaches.

Within the two decades after World War II, with the rapid development of modern engineering technologies and systems, particularly with new materials being invented, two problems with deterministic approaches became more and more obvious: firstly, specific information on the material strength and the load were not always available so that the applied deterministic approach may not have been correct; secondly, even if they were, such an approach would not be economical as the reserved safety margin, the difference between the strength of the material and the anticipated stress, was often overestimated [6]. Probabilistic approaches were then reconsidered. Although there was an argument that the distribution information was not always available either for some kinds of material property or load effect, it was unwise not to use them if they were available or could be obtained with little effort [8]. Uncertainties should not be ignored or suppressed by *safety factors*. In the last 40 years, with the help of the increasing computational capability, many non-deterministic methods have been developed and adopted in structural design practice, though some of them may not have been initially developed for analysis of engineering systems. Successful examples include the Monte Carlo simulation method (MCS), FORM (first-order reliability methods)/SORM (second-order reliability methods), asymptotic methods, and response surface methods (RSM). Recently, some non-probabilistic methods, or *possibilistic methods*, have been proposed and studied, such as the *interval*, *convex*, and *fuzzy methods*.

Modern non-deterministic reliability approaches were initiated by Mayer and Khozialov [6–8] and contributions were made by many others, including Freudenthal, Cornel, Ditlevsen, Lind, Moses, Rackwitz, Rosenblueth, and Schueller [6,8]. Today, probabilistic methods have been proven to be extremely useful in the analysis of many engineering problems such as flexible buildings subjected to earthquakes or wind, offshore structures subjected to random wave loading, aircraft structures undergoing fatigue failure, and space structures subjected to environmental temperatures [6]. The study of the possibilistic approaches, however, still remains at an academic level, and there are currently ongoing research projects to combine the two approaches [9]. The history of the development of structural analysis methods is summarized in Figure 1.4.

1.2 NON-DETERMINISTIC RELIABILITY ANALYSIS METHODS

As stated earlier, all engineering design problems involve uncertainties to varying degrees. A typical example in practice could be a beam structure with its length explained in different ways, depending on the information available and the nature of uncertainty. For example, "it has a mean value of 2.5 m and a standard deviation of 0.05 m, and it follows a normal distribution", "it lies between 2.4 and 2.6 m," or "it is about 2.5 m". The analysis of random problems can be treated by a probabilistic method, an anti-optimization method, or a method based on fuzzy theory.

As illustrated in Figure 1.5, non-deterministic approaches include probabilistic approaches and possibilistic approaches. There are two basic and widely used probabilistic approaches, namely the Monte Carlo simulation (MCS) method and the first-order reliability method (FORM). There are two main possibilistic approaches that are distinguished by the way the uncertainties are described. These approaches are *interval analysis* and *fuzzy theory*. There is another popular method called the *response surface method (RSM)*, which is used to analytically approximate the unknown safety margin, followed by the FORM, MCS or a possibilistic method.

This section presents a general introduction of the development of these methods, while the applications of these methods to dynamic systems will be reviewed in the next section. The detailed technical background of the two probabilistic approaches and the RSM, on which the present work is focused, will be given in Chapter 2.

1.2.1 Monte Carlo Simulation (MCS) Method

The Monte Carlo simulation (MCS) method, sometimes called the Monte Carlo (MC) method, was developed to produce approximation solutions to a variety of mathematical problems where a set of probabilistic observations

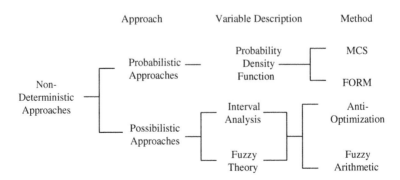

FIGURE 1.5 Categories of non-deterministic approaches.

is required. It performs a repetitive simulation process via random sampling. Though some earlier applications of the method were given more generic names such as "statistical sampling", the name "Monte Carlo" was finally arrived at to refer to the random number generator used in the famous casino in Monaco, which implies the randomness and the repetitive nature of the process. One of the early major contributors, Stanislaw Marcin Ulam, a Polish mathematician, tells in his autobiography that the method was named in honor of his uncle, who was a gambler [10]. Ulam suggested the Monte Carlo method for evaluating complicated mathematical integrals in the theory of nuclear chain reactions. This was whilst he was working on the Manhattan Project at Los Alamos during World War II and the postwar period. Many others, including Enrico Fermi, John von Neumann and Nicholas Metropolis, were among the early pioneers who contributed to the development of this simulation method [11].

However, since a large number of computations were required, it was only after the invention of electronic computers in 1945 that the Monte Carlo method began to be studied in depth. Apart from the early research conducted at Los Alamos, the Rand Corporation and the US Air Force were two major organizations responsible for funding research on the Monte Carlo method. Today, with increasing computational power, the Monte Carlo method has been widely applied in fields covering economics, physics, biology, traffic management, and engineering.

The underlying idea of Monte Carlo simulation is fairly simple. It employs techniques to select samples randomly to simulate a large number of experiments and observe the results [7,11,12].

For a large and complex engineering structure with uncertain properties and load information, the failure situation cannot be defined in a closed form. Instead, it is evaluated via a numerical analysis approach such as the finite element (FE) method [13−16]. The Monte Carlo simulation method is therefore an effective tool to handle this problem, particularly with the help of modern computational power [17]. There are a number of commercial software packages available now to conduct Monte Carlo simulations for various requirements.

However, the computational cost of Monte Carlo applications, by completing a large number of finite element executions, is often prohibitively high, particularly when many random variables are involved.

1.2.2 FORM (First-Order Reliability Method)

While the Monte Carlo method was not practised intensively due to computational limitations, some other probabilistic methods were developed. One of them was the first-order reliability method or FORM, which is widely used due to its simplicity, efficiency, and accuracy advantages [13,18,19]. It is given its name because it is a linear (first-order Taylor series) approximation solution involving statistical information about the random variables. The probability of failure finally obtained by the FORM method is usually quite accurate,[4] though approximate.

In the earlier example, if the strength resistance variable R and the load effect variable S are normally distributed, the margin of the two variables, where R exceeds S, is also a normally distributed variable, denoted as M, as illustrated in Figure 1.6. The mean and standard deviation of the margin variable M can be determined by the mean and standard deviation of R and S respectively. Obviously, the probability of failure is the probability that M is smaller than zero, shown in the figure as the shaded area represented by P_f. *Beta* (β) is the so-called *safety index*, which, in this one random variable example, defines the shortest distance between the mean and zero in terms of the standard deviation (σ). Obviously, the greater the value of β, the further the failure line (0) will be from the mean (the smaller the failure area will be), and therefore the lower the probability of failure P_f will be. Details of the definition of β in terms of more variables will be discussed in Chapter 2.

Basler and Cornell developed a framework of the FORM method that can be easily developed into a design code [20]. A significant drawback called the *invariance problem*, however, was found later by Ditlevsen [21], which results in the method possibly delivering different outcomes for the same engineering problem if the safety margins are expressed differently. This problem was finally solved by Lind and Hasofer in 1974 [22] by standardizing the variables and reconstructing the safety margin in the standard space.

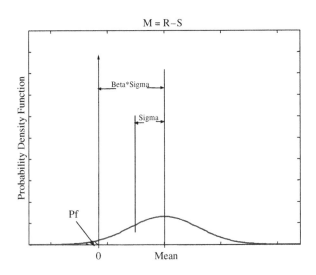

FIGURE 1.6 Example of the FORM method with one random variable.

---

4. However, this is for static systems only. Further discussion will be presented in the following section and Chapter 2.

To enhance the performance of FORM, two major steps were made thereafter. One is the development of the Rackwitz−Fiessler (R-F) method in 1978 [23], which deals with non-Gaussian problems, another is the Advanced FORM (AFORM) method or H-R method developed in 1981 by Hohenbichler and Rackwitz [24]. The H-R method adopts R-F techniques as well as addressing the problems of dependency between the random variables involved.

1.2.3 Interval Analysis

In interval analysis, the uncertain parameter is denoted by a simple range marked by lower and upper bounds. There is no distribution density information available for the variable. The aim of the method is to derive bounds on the response quantities of interest [9,25].

As illustrated in Figure 1.7(a), a single random variable is described between the two boundaries along its continuous linear domain. The value of the variable can be any within the range. The dotted curved line is a reference probability density distribution (a normal one in this example) of the variable that could be used to describe the variable in probabilistic methods.[5] The interval method for two parameters and three parameters can be extended to square and cube domains in Figure 1.7(b) and (c) respectively.

The interval arithmetic, on which the recent research work and development is mainly based, was introduced by Moore [26]. He developed interval vectors and matrices and the first nontrivial application. Alefeld and Herzberger's book, published in 1983, presents fundamental concepts of the interval algebra [27]. Contributions were also made by Neumaier in 1990 [28] and Hansen in 1992 [29]. Elishakoff is credited with introducing the use of interval analysis in engineering problems with uncertainties [30].

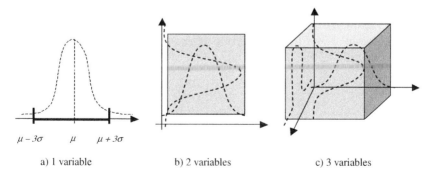

a) 1 variable   b) 2 variables   c) 3 variables

FIGURE 1.7   Bounded input variables.

---

5. $\mu - 3\sigma$ and $\mu + 3\sigma$ of pdf can be marked as the lower bound and the upper bound respectively in the interval method.

Combined with the finite element method, the interval analysis method can be applied to structural problems where only the bounded information of the property and load variables is available [31,32]. For example, the elements of the stiffness matrix and the components of the force vector with uncertainties can then be defined in bounded intervals and the relevant interval algebra calculation can be applied.

However, a major problem of interval analysis is the overestimation of the solution bounds due to the rounding (approximation) error during numerical analysis, resulting in the width of the interval functions growing with the number of interval variables and the number of computations involved [32,33]. Many new techniques are found in the literature to address this issue, which will be discussed in Section 1.3.2.

1.2.4 Fuzzy Analysis

In fuzzy theory, a fuzzy variable is defined as a member of a fuzzy subset of the variable domain. The degree of such membership is described by a degree of membership represented by the *membership function*, with the outcome value between 0 and 1 inclusive, denoted as α in Figure 1.8. "1" means a definite member of the subset and "0" means definitely not a member. For any other value, the membership is uncertain, which is in contrast to the classical set where membership and non-membership are clearly distinct. Therefore, the membership function is considered as a *possibility distribution function*, defining the intermediate possibility between strictly impossible and strictly possible events. As there is no associated statistical data, the choice of such a possibility distribution of a quantity is normally based on experts' opinion [25].

One type of fuzzy class, called *normal fuzzy numbers*, is widely used to represent the numerical uncertainties. In terms of the membership functions of the normal fuzzy members, as illustrated in Figure 1.8, there is at least one point where the membership is equal to 1 and the membership for the left side of the point is increasing while for the right side it is decreasing.

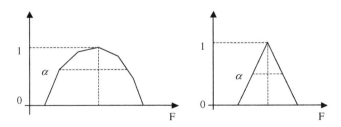

FIGURE 1.8   Membership functions of normal fuzzy numbers.

Zadeh introduced the theory of fuzzy sets in 1965 [34], and also extended it to be a basis of reasoning with possibility [35]. The concept of fuzzy logic was invented in 1967 and has since gained, particularly in the last two decades, an increasing popularity in practical applications, such as human-like decision-making processes in controller design that requires rules based on accumulated expert knowledge rather than strict objective data [36].

In conjunction with finite element analysis, fuzzy logic analysis can be used to handle the problems of structural uncertainties. This conjunction initiated the development of the fuzzy finite element method (FFEM). The aim is to obtain the fuzzy description of the system response. Typically, the FFEM consists of three major processes. (1) Definition mode: to define all the variables with uncertainties to be represented in fuzzy format (with membership functions), such as material/geometry properties and the constraints and loads. (2) Assembly mode: to assemble all the fuzzy system matrices. (3) Calculation mode: to obtain analysis results according to the fuzzy theory techniques.

The application of fuzzy theory has attracted significant interest in the engineering field and appropriate treatments can be found where there are vaguely defined system characteristics, e.g. imprecision of data, insufficient information, and subjectivity of opinion or judgment [37−43].

1.2.5 Response Surface Method (RSM)

Initially developed in the chemical processing industry in 1950s, the response surface method (RSM) is used to uncover analytically an unknown relationship (*an empirical model*) between several inputs and one output. The input variables are called *independent variables* (or *regressors*) and the output measure is called the *response* [44]. Usually, such a relationship is expressed in a *first-order* or *second-order* polynomial function called a *response surface model*. The *linear regression approach* is then applied to fit the response surface model and the model-fitting process may be repeated several times according to some kinds of convergence criteria until a satisfactory response model is obtained.

The advantage of RSM in uncovering an unknown relationship efficiently makes it ideal for engineering reliability analysis to approximate the unknown safety margin expression. However, the method was not applied to reliability analysis in the engineering field until 20 years ago, when sufficient computational power became available [45]. Since then, the response surface method has become an effective modeling technique that simplifies the computationally costly FE process (of MCS). RSM fits a polynomial model, i.e. a response surface model, after a limited number of FE runs, to approximately replace the implicit safety margin function.

Once a satisfactory response surface model is developed, which is often in a compact first/second-order polynomial form, some reliability evaluation techniques, such as FORM, SORM, the MCS method or the fuzzy method [43], can then be applied to obtain the required probability of failure.

Most applications of RSM are sequential, which implies that the whole experiment can be processed in some standard steps. This is particularly applicable to computer software. Three fundamental phases are summarized by Myers and Montgomery in their book [44]. (1) A *screening experiment*, in which the important variables (factors) will be identified while the unimportant ones will be eliminated. (2) *Optimum identification*, in which a set of independent variables are determined that results in a value of response that is near the optimum. A first-order model is most often used in this phase and the dominant optimization technique employed is called the method of *steepest ascent*. (3) A *true response approximation*, in which a refined model will be found to approximate the true response within a relatively small region around the optimum. A second-order model or a higher-order polynomial is often needed to fit the curvature at this stage.

In addition to Myers and Montgomery's book, which is state of the art, there have been two other textbooks published specifically on the response surface methodology,[6] by Box and Draper [46] and Khuri and Cornell [47]. Some other review papers have been published on this subject, such as Myers et al. [48], Hill and Hunter [49], and Myers [45]. These books and papers present fundamental ideas and techniques of the response surface methodology, as well as depicting its development history, modern applications, and future directions.

In the last decade, a number of papers were published in the engineering field with specific interest of applying the response surface method to reliability analysis. These in-depth studies cover a wide range of applications from iteration optimization [50−57], sampling design methods and optimization [58−64], to model adequacy testing [65−68] and applications [69−72]. These approaches were discussed in detail in Ref. [73].

1.2.6 Summary

The choice of one of the probabilistic or possibilistic methods depends on whether the information available for uncertain variables is either in the form of bounded values, degrees of membership, or probability density. However, in contrast to their distinct contents, there are strong connections between these methods, for example, the three-sigma bounds in probabilistic

---

6. The response surface methodology is a collection of statistical and mathematical techniques that are applied to fit and improve some predefined mathematical models, and to optimize the response according to a requirement [44]. It integrates the knowledge of sampling design techniques, regression modeling techniques, and elementary optimization methods.

methods are assumed to be realistic interval bounds for the interval analysis method [25]. If a function was used to describe the degrees of choosing different values within the range in the interval analysis method, fuzzy theory could be applied. If such a function were in the form of a description of a random density distribution, probabilistic methods would be appropriate. The response surface method can be applied in connection with either probabilistic or possibilistic methods.

There are always debates in the literature about whether to distinguish between the probabilistic and possibilistic approaches. On one side, some researchers believe that the probabilistic approach is only a subcategory of a more universal non-probabilistic approach. This would represent a more unified approach for non-deterministic analysis. Others argue that probabilistic methods are able to model anything the non-probabilistic approach can model [25]. Current studies and projects can be found in the literature attempting to combine these two approaches [9].

1.3 UNCERTAINTY ANALYSIS OF DYNAMIC SYSTEMS

1.3.1 Background

One of the great challenges for reliability analysis methods is to efficiently and accurately predict the reliability of *dynamic engineering systems*, in addition to the constant design demand for higher levels of safety, speed, conformability and economy, as typically required in the civil aviation industry. The difficulty is partly due to the complexities and diversities of the systems themselves and their working environments. These systems can be from everyday life transportation vehicles to some special giants such as offshore structures and space shuttles. Figure 1.9 presents an example set of engineering systems categorized by their excitation and response frequencies, from 1 to 20,000 Hz.

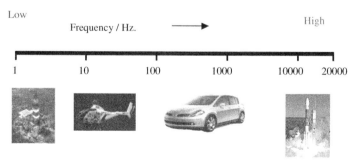

FEA – Finite Elements Analysis   SEA – Statistical Energy Analysis

FIGURE 1.9 Structural dynamics categorized by frequency.

A structure's dynamic characteristics are represented by its *natural (resonant) frequencies, damping values,* and *mode shapes.*[7] These characteristics are fundamental to vibration and noise, which are often the major operational considerations for designers and users [77,78]. Another important feature is the *modal overlap factor,* which marks the degree of overlap of responses between neighboring natural frequencies. The factor is low in the low-frequency range where there is a distinct gap between natural frequencies, while it is high in the high-frequency range where responses overlap.

The different phenomena between the dynamic problems of low frequency and high frequency have determined the different requirements of analysis techniques, which are illustrated in Figure 1.10.

Finite element analysis (FEA) is the most common tool used to analyze the performance of dynamic systems, particularly for low-frequency range systems. In the FEA, a structural model is defined in a finite element program so that the dynamic can be solved through numerical techniques and the dynamic characteristics can be obtained by *eigen solvers.* The detailed uncertainties of the system's random parameters can be described by a *parametric* form, which includes probabilistic approaches, e.g. Monte Carlo simulations, as well as possibilistic techniques.

However, it becomes more difficult to analyze the dynamic system using FEA at higher frequencies due to the involvement of many degrees of freedom and many uncertainties of model parameters. Instead, the problem can be approached using statistical energy analysis (SEA), which is an approximation tool using an average energy solution. While being disadvantageous

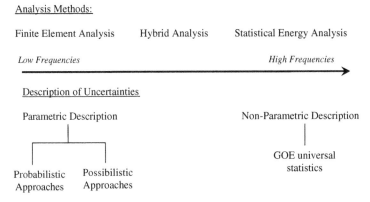

FIGURE 1.10 Different analysis requirements between low frequency and high frequency.

7. The fundamental characteristics of a real dynamic system can be determined and analyzed through a set of experimental and theoretical procedures called *modal analysis.* It consists of measurements of a series of frequency response functions (FRFs) at various geometric locations, and mathematical analysis approaches to give an overall understanding [77].

for FEA, the high randomness in the high-frequency range gives opportunities to other analytical techniques that are not dependent on the uncertainty of the system's random parameters. Recent studies have shown that, at high frequencies and given that the system is random enough, the statistics of the response, which is a result of the statistics of the eigenvalues and eigenvectors, are independent of the detailed nature of the uncertainties of the system parameters [79]. The uncertainties at high frequencies can therefore be described in a *non-parametric* form. A universal statistics technique, called the Gaussian orthogonal ensemble (GOE), can be applied to predict the variance of the response. By exploiting universality and GOE statistics in mid- to high-frequency problems, significant progress has been made using SEA and the *hybrid* method (which combines the FEA and SEA) [79,80].[8]

For reliability analysis of a low-frequency problem, both probabilistic and possibilistic approaches can be applied to deal with different kinds of uncertainties. The most used probabilistic approach is the MCS method. However, due to the considerable computational cost, there is a severe limitation on the utility of the MCS method, particularly for large dynamic systems with many random variables such as aircraft wings. Possibilistic approaches are also studied for structures with uncertain but non-random parameters involving a substantial number of engineering vibration problems [82]. In fact, within the last decade, there is increasing research interest in applying the interval and fuzzy methods when little statistical information on uncertainties of the system is available [25]. The approaches and techniques for the analysis of dynamic systems in the literature are reviewed and presented in the next section, and a summary of the associated applications is given in Table 1.1.

1.3.2 Literature Review of Analytical Approaches to Dynamic Systems

Interval Analysis

Most of the examples of interval analysis in the literature involve a first-order perturbation matrix approach and are concerned only with the interval eigenvalues.

Qiu et al. [83−85] presented an approximation method based on matrix perturbation theory to evaluate interval eigenvalues of non-random interval stiffness and mass matrices. The interval stiffness and mass matrices ($\mathbf{K}^I$, $\mathbf{M}^I$) are decomposed into the nominal matrices ($\mathbf{K}^c$, $\mathbf{M}^c$) and the deviation amplitude matrices ([$-\Delta K, \Delta K$] and [$-\Delta M, \Delta M$]). Therefore, the generalized eigenvalue problem can be replaced by the perturbation form. The matrix perturbation

---

8. At high frequencies, the response "variance" is calculated and the distribution (pdf) of the response can be determined (such as log-normal distribution [81]). However, this is not the case for the low frequencies, where a known "variance" alone is not adequate to predict the response.

TABLE 1.1 Summary of All the Reviewed Papers

Reference	Method	Deliveries			Numerical Examples		
		Eigenvalues/ Eigenvectors	FRFs	Others	Structural Model	No. of DOFs	No. of Uncertainties
Qiu, Chen and Elishakoff (1996) [83]	Interval analysis	√			4-mass/5-spring system	4	8
Qiu, Chen and Elishakoff (1996) [84]	Interval analysis	√			15-bar planar frame	13	6
Qiu, Elishakoff and Starnes (1996) [85]	Interval analysis	√			5-story frame	5	10
					32-bar truss	32	4
Chen, Zhang and Chen (2006) [87]	Interval analysis	√			Cantilever plate	300	36
					5-story frame	5	5
Dimarogonas (1995) [82]	Interval analysis	√		Interval response	Turbine rotor	3	3
Dessombz et al. (2001) [32]	Interval analysis		√		2-mass/3-spring	2	3
					Clamped free plate	20	1
					3-bladed-disk	7	1
Chen and Wu (2004) [88]	Interval analysis			Interval response	Truss structure	30	3
					Frame structure	30	2
Chen and Rao (1997) [39]	Fuzzy finite element	√			3-stepped bar	3	12
					25-bar space truss	18	10

method was employed to solve the generalized interval eigenvalue problem, and the bounded eigenvalues were approximately determined.

By applying the above perturbation method to some numerical examples (of dynamic systems) and comparing the results to Deif's method [86], this perturbation method was found to be much more efficient, though larger interval eigenvalues were generated. It was also seen that the higher the order of the eigenvalues (natural frequencies), the smaller the influence of the interval uncertainties on eigenvalues.

Chen et al. [87] extended the matrix perturbation and interval extension theory to damped dynamic systems. A method for estimating the upper and lower bounds of the complex eigenvalues of closed-loop systems was developed. The results were derived in terms of the eigenvalues and left and right eigenvectors of the second-order system.

The applicability of the method for FE models and the validity and efficiency of the method were tested using two numerical examples. The following conclusions were drawn:

1. The effect of changes of the input internal parameters on the imaginary parts of the eigenvalues is larger than that on the real parts. The results are useful to estimate the stability robustness of a control system with uncertainties.
2. The methods for constructing the interval stiffness, mass, and damping matrices were presented by using the physical parameters to avoid using bounds in Euclidean norm for uncertain matrices.
3. The method is easy to implement on a computer and easy to incorporate into the finite element code. Therefore it can be applied to large finite element models.
4. The computational efficiency of the interval method is higher than that of the classical random perturbation method, which takes $\pm 3\sigma$ as the upper and lower bounds.

Dimarogonas [82] claimed that the common interval techniques could not be directly applied to find dynamic responses of a linear system with interval matrices of system parameters, due to the interval divergence problem. Instead, the interval modal analysis and the interval solution of the eigenvalue problem were developed to predict the response of a rotor dynamic system. If the eigenvectors do not change their signs, the exact intervals of the eigenvalues can be determined. Otherwise, a subdividing procedure is run where the interval of each interval parameter is divided into two equal sub-intervals, and the signs of the eigenvectors in each sub-interval solution are examined. 2^n vertex interval eigenvalue solutions are then required. The procedure continues if there is a sign change and the sub-intervals are halved again, until convergence is obtained. Obviously this method can be computationally intensive if there are many interval parameters so that a large number of eigenvalue solutions may be needed. However, it was argued that this is seldom the case for engineering systems, because the number of interval quantities is relatively small.

A rotor dynamic example was presented and a Monte Carlo method was developed to provide the exact interval solution to compare with that of the proposed method. The eigenvalues were calculated by the two methods and a very good agreement was found.

Dessombz et al. [32] performed an interval calculation method to find an envelope of transfer functions for mechanical systems modeled by finite elements with uncertain and bounded parameters. A new iterative algorithm was developed to avoid the overestimation problem. The algorithm was based on the iterative fixed-point theorem with a splitting convergence technique. If the iteration matrix is not contracting, the convergence will be guaranteed by a dichotomy scheme that splits the centered interval matrix into sub-intervals until the algorithm converges. The faster the convergence of the algorithm, the smaller the overestimation of the solution will be.

Several numerical examples were studied. The real and imaginary parts of the transfer function were calculated by the proposed algorithm and compared to the results obtained by a Monte Carlo simulation method. It was seen that the envelope produced by the new algorithm wrapped that of the deterministic transfer functions determined by the Monte Carlo method, with only small overestimations.

Chen and Wu [88] presented an interval optimization method for analyzing the dynamic responses of structures with interval parameters. The method combines the interval extension technique with the perturbation theory to predict the interval dynamic responses. The interval optimization problem was simplified to an approximate deterministic optimization by performing a first-order Taylor expansion of the interval objective function.

The optimization method was applied to a truss structure and a frame structure. Using the Lagrange optimal algorithm, the approximate linear deterministic optimization problem was solved to reduce the displacement amplitude, subject to the constraint conditions involving the interval optimization parameters. The advantage of the method is that more information for optimal structures can be obtained, such as how the optimization results change with regard to the change of the uncertainties of structural parameters. It was found that, when more variables were taken as the optimization parameters, the interval value of the objective function obtained became sharper.

However, because the approach is based on the first-order Taylor expansion, the application of the approach is limited to cases where the interval uncertainties of the parameters are small. For larger uncertainties, a second-order Taylor expansion is needed.

Fuzzy Finite Element Analysis

Chen and Rao [39] presented a computationally efficient method for fuzzy finite element (FFE) analysis of dynamic systems involving vagueness. The fuzzy eigenvalue and eigenvector problems were solved using the concept of interval analysis with the input of fuzzy stiffness and mass matrices. An efficient approach was developed to handle the nested interval operations at each α level by using Taguchi's robust philosophy, with some modifications of the normalization. Two-level fractional factorial design (with two-level orthogonal arrays) was employed to reduce the total computational effort.

Two numerical examples were performed and the fuzzy frequencies and fuzzy mode shapes were calculated. The accuracy of the solution and the effort required to find the solution depend on the initial value of the search step length. The smaller the initial step length, the more accurate the solution will be.

This work was claimed to be, in the engineering field, the first attempt at fuzzy finite element analysis involving dynamics. The proposed approaches, however, need to be extended for finite element analysis of (imprecisely defined) fuzzy engineering systems involving dynamic forces.

Wasfy and Noor [89] developed a computational procedure for predicting the dynamic characteristics of flexible multi-body systems with fuzzy parameters. This fuzzy method has been used to find time-histories of the possibility distributions of the response and sensitivity coefficients that measure the sensitivity of the dynamic response to variations in the material, geometric and external force parameters of the system.

Key techniques applied in the method involve a total Lagrangian formulation, a vertex method, a semi-explicit temporal integration technique and a direct differentiation approach. The effectiveness of the method and the usefulness of the fuzzy output were demonstrated using three numerical examples, including an articulated space structure consisting of beams, shells, and revolute joints. At most, three fuzzy parameters were chosen in each problem, and all were assumed to have triangular membership functions. Three α-cuts (0, 0.5, and 1.0 respectively) were used and a total of 17 crisp computations were needed. Fuzzy time envelopes of the desired response quantity for each α-cut were calculated using the vertex method. The envelopes of the strain energy, total energy, responses and sensitivity coefficients of the strain energy and total energy, with respect to the change of the fuzzy parameters, were presented and discussed.

Lallemand et al. [90] applied the fuzzy set theory to dynamic finite element analysis of engineering systems with uncertain material properties. A general algorithm was developed to process the uncertain eigenvalue problem. The method was extended from a specific algorithm based on the perturbation modal analysis. The techniques of fuzzy entropy, specificity, and defuzzification were available for measuring the imprecision of the uncertain system parameters.

A plate structure was studied. Three material parameters, Young's modulus E, Poisson's ratio ν and material density ρ, were selected as fuzzy input parameters. Fuzzy frequencies were calculated by the method according to the simultaneous perturbations of E, ν and ρ. It showed that the modal behavior was more affected by parameters E and ν rather than ρ. Based on the studies of the entropy and specificity of the fuzzy frequencies, it was concluded that the more extensive the entropy, the less certain the frequency's value will be.

In contrast, the more extensive the specificity, the more deterministic the frequency value will be.

Moens, Vandepitte and colleagues [91–97] implemented the fuzzy finite element technique as an alternative to the Monte Carlo simulation method, for the analysis of structural dynamics. The classical fuzzy method was extended to a more realistic generalized FFE approach. The approach aims to provide a fuzzy description of the envelope of the frequency response function (FRF) with fuzzy uncertainties on the input model, using the FFE methodology and the modal superposition principle.

The effectiveness and efficiency of the proposed approach were demonstrated using a number of numerical examples. The input parameters were described by the interval and the fuzzy α-level technique. Triangular membership functions were used in all the examples. The modal envelope of the FRF was calculated using the modal superposition algorithm. The calculation had to be taken on each mode. However, it could only involve some selected modes that contribute to the total FRF in the frequency domain of interest.

The corner-eigenvalue method and the optimized vertex method were performed and compared to the corner method. As a result of the comparison, the corner-eigenvalue method was selected for the future investigation as it delivered the best results [94,95]. Advanced techniques were also applied to tackle the coupling problems of modal parameters [93,97].

Donders et al. [98] proposed the short transformation method (STM) as an alternative to the original transformation method (TM) to predict the FRFs of engineering systems with uncertain inputs. The technique involves a fuzzy approach based on interval arithmetic (an extension of the vertex method): the uncertain response is reconstructed from a set of deterministic responses, combining the extrema of each interval in every possible way.

Both methods were demonstrated on a car front cradle with uncertain design parameters. It was concluded that STM is more efficient than TM, because for n uncertain parameters and m interval levels, STM can yield a reduction of $2^n(m-1)$. This means that the STM allows reconstruction of the fuzzy FRFs from a much smaller number of deterministic computations, though there is a slight reduction in the predicting accuracy of FRFs.

Moens and Vandepitte [25] presented an in-depth review of the nonprobabilistic approaches for non-deterministic dynamic analysis. Firstly, a classification of different types of non-deterministic properties was presented and discussed. Secondly, an overview of possible implementations of the non-probabilistic finite element procedure was presented. Finally, a numerical example was presented to illustrate the applications of interval and fuzzy FRF analysis.

Monte Carlo Simulation

To reduce the computational costs of the applications of Monte Carlo simulation methods, advanced techniques are being sought and studied.

Pellissetti et al. [99] recently developed a class of advanced Monte Carlo simulation methods to efficiently analyze the reliability of large-scale structures modeled by very large FE systems. The computational costs were dramatically reduced as shown by the given numerical example, a spacecraft structure under dynamic loading, which was analyzed by a large FE model with 120,000 degrees of freedom (DOFs) and 1300 uncertainties.

Although it was claimed that the reliability analysis of a large FE model became feasible (based on this recently developed simulation approach), the efficiency and accuracy of the approach still depends heavily on the system's frequency range. The underlying reason is the strong nonlinearity of the limit state function in the space of the uncertain input parameters within the frequency range of interest. An ongoing benchmark study on reliability analysis of structural dynamics problems [28] has led to similar observations, i.e. "significant deterioration of the performance of virtually all methods in the presence of strongly nonlinear limit state functions". Therefore, the major difficulties are not due to the large number of parameters but the nonlinearities.

It is believed that substantial additional work is needed on this type of problem in general.

Response Surface Method

Margetson [100,101] has reported a failed attempt to apply the response surface method to directly predict the dynamic response of a forced plate structure with geometry and material random variables. Instead, the natural frequencies and a defined modal parameter were expressed as a function of the input random variables by a second-order response surface model. The desired displacement and stress responses were obtained by the modal superposition principle. The details will be discussed in the next chapter.

Stochastic Reduced Basis Method

Nair and Keane [102] presented stochastic reduced basis methods (SRBMs) for solving large-scale linear random algebraic sets of equations, such as those arising from discretization of linear stochastic PDEs in space, time, and the random dimension. The fundamental idea was to represent the response process using a linear combination of stochastic basis vectors with undetermined coefficients. These coefficients are either considered as deterministic scalars or random functions. The number of undetermined coefficients should be less than the dimension of the discretized PDE, for which the name "stochastic reduced basis methods" was given.

Efficient procedures for computing the basis vectors (with a particular emphasis on the preconditioned stochastic Krylov subspace) were presented, particularly for stochastic structural dynamic analysis. Two variants of the Bubnov−Galerkin (BG) scheme were employed to compute the undetermined coefficients in the stochastic reduced basis representation: one was treated as deterministic scalars and the other as random functions. Both these projection schemes allow explicit expressions for the response quantities to be derived, which make it possible to efficiently compute the complete probabilistic description.

Numerical examples were studied and presented to demonstrate that high-quality approximations of the response statistics can be achieved for coping with large coefficients of variation of the random system parameters. The results calculated by SRBMs were compared to those of the Neumann expansion scheme and the polynomial chaos scheme, showing that the SRBM results are more accurate with a much lower computational cost.

Table 1.1 presents a summary of all the reviewed papers with application details.

1.3.3 Summary

Low-frequency range problems have attracted substantial research interest. It can be seen from the literature that various techniques have been investigated and applied to dynamic systems to deal with different kinds of uncertainties.

However, it can be seen that, at least at the academic level, there has been an unbalanced development of research interest on the probabilistic and possibilistic methods. In-depth studies have been carried out on applications of the possibilistic methods to dynamic systems. Effective and efficient methods have been developed and application examples involve some large and complex structures. In contrast, studies of the non-possibilistic methods have lagged behind. Nonlinearity of the limit state function is the major problem for almost all probabilistic methods [99]. As a result, FORM is not directly applicable to random vibration analysis [103,104]. This problem will be discussed in detail in Chapter 2. The strong nonlinearity can cause problems in using the Monte Carlo simulation method as well, in addition to the long-term problem of unaffordable computational costs. It was also reported that the response surface method is unable to accurately predict the dynamic response, which will also be discussed in Chapter 2. Problems of applying these methods are briefly summarized in Table 1.2.

TABLE 1.2 Problems in Applying MCS, FORM, and RSM Methods to Dynamic Systems

Method	Problem in Applying to Dynamic Systems
MCS	High computational cost and inaccuracy if nonlinearity of limit state function is present
FORM	Multiply connected and non-monotonic failure surface
RSM	Poor prediction of the dynamic response

1.4 SCOPE OF THE PRESENT WORK

The literature review has shown that more work is needed to deliver an efficient and accurate probabilistic approach to dynamic systems. The problems considered and presented in this book are:

1. Nonlinearity of the limit state functions
2. Intensive computational cost
3. Complexity of the dynamic systems.

A perturbation approach, based on an optimized modal model, is proposed to overcome the above problems, followed by use of one of the probabilistic methods to perform the reliability analysis of a dynamic system. Since the resonance cases are of most concern, the optimized model simplifies the complexity of the dynamic systems by only concentrating on the resonance dominating terms in the response element (expressed in terms of modal coordinates). This optimization and later newly defined parameters (replacing the original random variables) transform the original safety margin into an approximate (constructed upon the new parameters) but smooth and linear one. Finally, the statistical information of the new parameters can be derived from that of the original variables by solving just once the eigen problem on the mean values of the original variables. An efficient reliability method, such as FORM, can then be applied. This combined approach will be discussed and demonstrated in the remainder of this book.

In summary, the present research focuses on:

- dynamic systems
- low excitation frequency range
- probabilistic reliability methods.

The aim is to deliver an efficient reliability approach for engineering systems under dynamic loads.

1.5 OVERVIEW OF THE BOOK

- Chapter 2 introduces the technical background on the reliability analysis and the relevant probabilistic approaches, i.e. MCS, RSM, and FORM. The problems of applying these approaches to dynamic systems are discussed.
- Chapter 3 presents the fundamentals of the novel perturbation approach, which is the core of the proposed combined approach, and the derivation of the statistic moments (covariance matrices) of the newly designed parameters.
- Chapter 4 demonstrates an application of the combined approach, i.e. perturbation approach + FORM method, to a 2D framework structure with random added masses. The detailed analysis, including inaccuracy

problems, is reported and discussed. An alternative solution, i.e. the *perturbation approach + MCS*, is then determined.

- Chapter 5 presents the application of the combined approach to a 3D complex helicopter model. Different reliability analysis cases are performed and reported. The consistency and stability of the approach are demonstrated.
- Chapter 6 presents an implementation of the response surface method to derive covariance information of the stiffness matrix from that of the original property random variables. This is followed by a successful combined approach, which together with the covariance information offers a complete solution for reliability analysis of dynamic systems. The reliability analysis cases performed in Chapters 4 and 5 are re-run and the results are discussed.
- Chapter 7 draws conclusions and proposes future enhancements of the work.

Chapter 2

Technical Background

This chapter presents the technical background of the structural reliability issues and three methods of reliability analysis. The problems of applying these methods to dynamic systems are also discussed in detail. The solution, a modal perturbation-based combined approach, is then proposed.

2.1 DEFINITION OF STRUCTURAL RELIABILITY

There are always safety requirements for real-life engineering structures that must be satisfied. As discussed in the example in Chapter 1, a structural reliability problem can be described by the load effect S and the resistance R of a structural element. Both variables should be expressed in the same units and represented by a known probability density function (shown in Figure 1.2) to accommodate uncertainties. The structural element will fail if the load effect S exceeds its resistance R. Thus the *safety margin*, a mathematical expression representing the difference between the resistance and the load effect, is denoted as $M = R - S$. The safety requirement for this example is $S < R$ or $M > 0$. Such a requirement is called a *limit state* at which a threshold value of a specific response parameter should not be reached [7]. With uncertainties being involved, the evaluation of such a limit state is a probability outcome.

The probability of failure of the structural element is therefore defined as

$$P_f = P(R \leq S) = P(R - S \leq 0). \tag{2.1}$$

The safety design is to guarantee that either there is no violation of such state or the probability of such a violation is low enough.

The violation of a limit state of a structure depends on many issues including the magnitude of the applied load, the properties of the structure such as its strength and stiffness, and the boundary conditions.

In a complex structural problem, a more general term of n-dimensional vector of random variables, $\{X_i\}_{i=1}^n$, is collected to represent the characteristics of the applied load and those of the structural properties. The probability of failure can be calculated using the following equation [8]:

$$P_f = \int_{G(x) < 0} p_x(x) \mathrm{d}x, \tag{2.2}$$

where $p_x(x)$ is an n-dimensional *joint probability density function* of the random variables $\{X_i\}_{i=1}^n$, while $G(X) = 0$ is the *limit state function*, or the *failure surface*,[1] which separates the safe and unsafe regions in the space spanned by $\{X_i\}_{i=1}^n$, i.e. $G(X) < 0$ and $G(X) > 0$ respectively. In the above case, $G(R,S) = M = R - S$.

Structural reliability (r) is therefore a quantitative concept in terms of the probability of failure of the structure in question. It is defined as the complement of the probability of failure, i.e.

$$r = 1 - P_f. \tag{2.3}$$

The aim of *reliability analysis* is to calculate and predict this probability of failure using effective techniques available with the relevant information of the structural properties and the environment.

However, explicit evaluation of equation (2.2) is not possible in most cases of engineering interest due to the fact that the random variables $\{X_i\}_{i=1}^n$ can be mutually dependent and non-Gaussian and complete knowledge of $p_x(x)$ is often unavailable, or the failure surface $G(X) = 0$ can be highly nonlinear and sometimes may not be explicitly obtainable except through a numerical solution such as a finite element code, or the number of random variables can be very large [8].

To deal with these difficulties, approximation techniques are employed to evaluate equation (2.2) in reliability analysis practice. As introduced in Chapter 1, two probabilistic methods are widely used, namely (1) Monte Carlo simulation (MCS), a simulation-based probabilistic method, and (2) first-order reliability method (FORM), a safety index-based probabilistic technique. In addition, the response surface method (RSM) can be applied to replace the implicit failure surface with a polynomial equation. The reliability analysis can then be performed by the MCS and FORM methods. The background of these methods will be discussed in the following sections.

2.2 TECHNICAL BASIS OF THE MONTE CARLO SIMULATION METHOD

The Monte Carlo simulation method is often applied to a reliability problem where equation (2.2) cannot be obtained in a closed form [7].

For reliability analysis of large and complex structures, the sample values of the vector $\{X_i\}_{i=1}^n$ can be chosen randomly and the response can be solved through a finite element (FE) method. According to the FE output, which is for example often a value of stress or displacement under external loads, the relevant limit state function $G(\hat{x}_i) = 0$ is then checked, where $\hat{x}_i$ is a set of realizations of $\{X_i\}_{i=1}^n$. The structure or structural element would have "failed" if the limit state was violated, i.e. $G(\hat{x}_i) \le 0$. This experiment is

---

1. *Failure surface* is a more often used term referring to a multidimensional space.

repeated many times to get the desired accuracy and hundreds of thousands of trials are not unusual.

If N trials have been completed, the probability of failure can then be evaluated approximately by

$$P_f \approx \frac{n(G(\hat{x}_i) \le 0)}{N},$$ (2.4)

where $n(G(\hat{x}_i) \le 0)$ denotes the number of trials that have $G(\hat{x}_i) \le 0$.

Obviously, the more trials that are conducted, the higher the accuracy of the P_f estimate that is achieved. Since it simulates the real cases, the reliability results obtained by the Monte Carlo simulation method are widely believed to be the most "accurate" results[2] and often used as an "exact" reference for other methods to compare with.

It should be noted that the accuracy of the Monte Carlo method does not depend on the geometry of the structure or the distribution features of the random variables, but on the number of trials and the value of the probability of failure.[3]

For the purpose of efficiency, it is very important to decide in advance the sample size N of the MCS, i.e. the minimum number of trials needed for the MCS result to be accurate enough. This is not, however, an easy task and depends on the nature of the problem [7].

According to Rubinstein [105], the *coefficient of variation* c of the estimator (P) of the probability of failure P_f, with N independent sample values, is given by

$$c = \sqrt{\frac{1 - P}{PN}}.$$ (2.5)

When P is small, equation (2.5) becomes

$$c \approx \sqrt{\frac{1}{PN}} \quad (P \ll 1).$$ (2.6)

Therefore, for a given small enough c, N can initially be evaluated by

$$N = \frac{1 - P}{Pc^2} \quad \text{or} \quad N = \frac{1}{Pc^2} \quad (P \ll 1).$$ (2.7)

Many earlier studies have tried to estimate the number of simulations required by a given confidence level C and the target probability, in the

---

2. Given that the FE model is a perfect representation of the physics.
3. Often, the lower that the probability, the more trials that are needed to maintain the accuracy.

reliability analysis case P_f [7,11,12,106]. On the basis of the central limit theorem, such a confidence interval is defined as [7]

$$P(-k\sigma < J - \mu < k\sigma) = C, \qquad (2.8)$$

where J is the estimated value of P_f, and μ and σ are the expected value and the standard deviation of J respectively. For example, for the confidence interval $C = 95\%$, $k = 1.96$, which can be verified from standard normal tables [7].

According to Shooman [107], σ can be approximated by the binomial parameters $\mu = P$ and $\sigma = (Pq/N)^{1/2}$, with $q = 1 - P$, provided that $NP \geq 5$ when $P \leq 0.5$. Equation (2.8) then becomes

$$P\left(-k\left(\frac{Pq}{N}\right)^{1/2} < J - P < k\left(\frac{Pq}{N}\right)^{1/2}\right) = C. \qquad (2.9)$$

By definition, the coefficient of variation of the estimate is $c = \sigma/\mu$; substituting $\mu = P$ and $\sigma = (Pq/N)^{1/2}$ into $c = \sigma/\mu$, equation (2.5) will be obtained.

The error between the actual value of J and the observed value is denoted by

$$\varepsilon = k\frac{\sigma}{\mu} = kc. \qquad (2.10)$$

Substituting equation (2.6) into equation (2.10) gives

$$\varepsilon = k\sqrt{\frac{1}{PN}} \quad (P \ll 1). \qquad (2.11)$$

Rearranging equation (2.11) results in

$$N \geq \left(\frac{k}{\varepsilon}\right)^2 \frac{1}{P}. \qquad (2.12)$$

However, there is a condition which is that for J to be Gaussian (based on the central limit theorem), PN (the number of "failures") must be Gaussian. It is therefore assumed that $PN \geq 10$, which gives $N \geq 10/P$. Considering equation (2.12), the determining expression of N should be:

$$N \geq \max\left\{\left(\frac{k}{\varepsilon}\right)^2 \frac{1}{P}, \frac{10}{P}\right\}. \qquad (2.13)$$

In summary, the estimate of P_f by Monte Carlo simulations is illustrated in Figure 2.1. The question for determining the sample size of MCS, N, is now: "How large should N be, to be confident to a percentage level C that the simulation result of P_f will fall in the range R?" C determines k (for instance, if $C = 95\%$, then $k = 1.96$, as in the earlier example), and R determines ε. Equation (2.13) is the answer to the question.

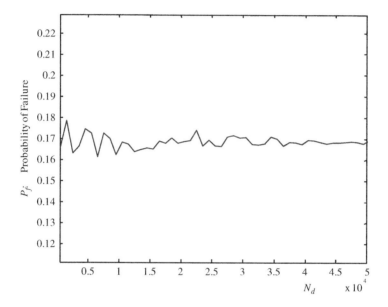

FIGURE 2.1 The true simulation result of P_f.

FIGURE 2.2 Plot analysis to determine the sample size of MCS.

For example, given an expected $P = 10^{-3}$, in order for the error in the estimate of P to be less than 20% with 95% confidence ($k = 1.96$, as in the earlier example), the required sample size N should not be less than 100,000. For another larger probability, $P = 10^{-1}$, N should be at least 1000.

Obviously, though often not practical, the most accurate approach to determine the sample size N is to run a series of MCS experiments with different values of N and carry out a plot analysis to observe the variation of P_f [7]. As illustrated in Figure 2.2, P_f should converge after N_d, which is the minimum sample size for the problem.

As a large number of simulations is sometimes necessary, the computational cost of completing the same number of finite element executions is often prohibitively high, particularly when many random variables are involved. To reduce the computational effort, a number of so-called *importance sampling* methods[4] have been developed, in which the realization of the trial can

---

4. More accurately, they should be called *weighted sampling* methods [11].

be focused on the failure region [7,11,12]. The total number of trials will be reduced while the accuracy is maintained.

2.3 THEORY OF THE FIRST-ORDER RELIABILITY METHOD (FORM)

The exact evaluation of equation (2.2) can be obtained if the safety margin is a linear function of the random variables $\{X_i\}_{i=1}^n$ and these variables are normally distributed.

In the example discussed earlier, supposing that the random variables S and R are independent and both normally distributed with means μ_R and μ_S and variances σ_R^2 and σ_S^2 respectively, the first two moments of the safety margin M (defined as $M = R - S$), i.e. mean μ_M and variance σ_M^2, can then be determined by

$$\mu_M = \mu_R - \mu_S$$
$$\sigma_M^2 = \sigma_R^2 + \sigma_S^2. \tag{2.14}$$

Referring to Figure 1.6, and using equation (2.14), equation (2.1) gives [7]

$$P_f = P(R - S \leq 0) = P(M \leq 0) = \Phi\left(-\frac{\mu_M}{\sigma_M}\right) = \Phi\left[-\frac{(\mu_R - \mu_S)}{\sqrt{\sigma_R^2 + \sigma_S^2}}\right], \tag{2.15}$$

where $\Phi(\ )$ is the cumulative distribution function for standard normal variables. The above equation can also be rewritten as

$$P_f = 1 - \Phi\left(\frac{\mu_M}{\sigma_M}\right) = 1 - \Phi\left[\frac{(\mu_R - \mu_S)}{\sqrt{\sigma_R^2 + \sigma_S^2}}\right]. \tag{2.16}$$

Cornell [20] denoted the ratio μ_M/σ_M as β, i.e.

$$\beta = \frac{\mu_M}{\sigma_M}, \tag{2.17}$$

and named it the *safety index*; therefore,

$$P_f = \Phi(-\beta) = 1 - \Phi(\beta). \tag{2.18}$$

The Cornell method can be generalized to many random variables by taking a Taylor series expansion of the limit state function about the mean values. Truncating the series at the linear terms, the first-order approximation mean and variance (standard deviation) can be obtained and used to evaluate equations (2.17) and (2.18).

However, when the limit state function is nonlinear, Cornell's safety index fails to be constant under different but mechanically equivalent formulations of the same limit state function. For example, the safety index for a set of mechanically equivalent limit state functions, namely $g_1(R,S) = R - S$, $g_2(R,S) = (R - S)^2$, and $g_3(R,S) = \log(R/S)$, would be expected to be identical, but they are not in Cornell's method. This so-called *invariance problem* was found by Ditlevsen [21], which indicates that the results of the method depended on how the limit state function was formulated. This problem was overcome by Hasofer and Lind [22], who found that β also represents the shortest distance from the origin to the failure surface when the surface is plotted in the space of standard normal random variables.

Therefore, the Hasofer and Lind (HL) method contains three steps: (1) making an orthogonal transformation of variables $\{X_i\}_{i=1}^n$ into a set of independent random variables (a de-correlation algorithm is presented in Appendix I); (2) introducing reduced variables and transforming the original limit state function into a standard space; (3) calculating the shortest distance from the origin to the new failure surface.

Returning to the earlier example, it is assumed now that the limit state function $G(R,S) = 0$ is nonlinear. To apply the HL method, the two random variables R and S (assuming they are correlated), denoted in more general terms, X_1, X_2, can be firstly transformed into independent variables Y_1, Y_2, and then be standardized into Z_1 and Z_2 respectively as

$$Z_1 = \frac{R - \mu_R}{\sigma_R} = \frac{Y_1 - \mu_1}{\sigma_1}, \quad Z_2 = \frac{S - \mu_S}{\sigma_S} = \frac{Y_2 - \mu_2}{\sigma_2}. \tag{2.19}$$

$G(R,S) = 0$ will then correspond to $G(Z_1,Z_2) = 0$ in the space of Z. The safety index can then be obtained by calculating the shortest distance from the origin to $G(Z_1,Z_2) = 0$.

Figure 2.3 presents the *geometrical interpretation* of the HL safety index $|\beta_{HL}|$, which is the shortest distance from the origin to the failure surface in the space of transformed variables Z. The point on the failure surface that lies closest to the origin is called the *design point*, also called the *most probable point* (MPP). At this point, the joint probability density is the greatest and the structure is most likely to fail [13]. Geometrical solutions of each point in terms of known moments are given in Figure 2.3. The shortest distance from the origin to the failure surface (O to MPP), β_{HL}, can then be evaluated, i.e. equation (2.15) is obtained.

It is clearly seen in the figure that there is a direct relationship between the safety index and the probability of failure. As $|\beta_{HL}|$ increases, the failure surface moves away from the origin and the failure region becomes smaller (as does the volume prescribed by the failure region and the joint probability density function, according to equation (2.2)), and thus the probability of failure decreases. Given that if the failure surface is a linear function, and R and S are independent random variables following a standard normal distribution, the HL safety index will be *exact*, as will $P_f = \Phi(-\beta_{HL})$. For a

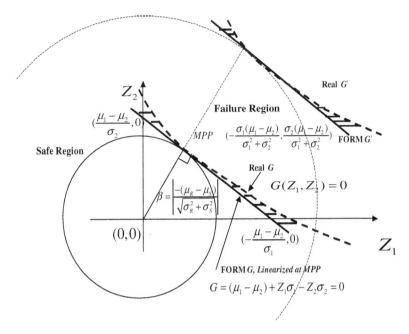

FIGURE 2.3 FORM method — a linear solution (on two random variables).

nonlinear failure surface, shown in the figure as the curve "Real G", the HL safety index is only an approximate result. The problem will be linearized at the design point, represented by the curve "FORM G" in the figure. The design point z^* is the tangent point of the failure surface $G(z) = 0$ interfacing with the hypersphere with a radius of β (see Figure 2.3 for a two-variable example).

The error between FORM G and the curve Real G is represented by the shaded area in the figure. Obviously, the smaller the area, i.e. the closer that linear FORM G is to the curve Real G, the more accurate the approximate FORM result would be. Considering a case of larger β (a smaller probability of failure), with the failure surface Real G' (same curvature as Real G) and the approximation FORM G' shown in Figure 2.3, the error area between the two is smaller than that between Real G and FORM G, which indicates that the FORM approximation for smaller P_f is better than that with larger P_f. Also, due to the fast $e^{-\lambda}$-decaying jointly normal probability density for smaller P_f values, the overall difference between the results of P_f (the volume prescribed by the probability density function and the failure region, according to equation (2.2)) would also be smaller. In conclusion, the FORM result would be more accurate in analyzing the reliability of a smaller P_f value than that of a larger one [7], given that other properties are the same (for example, curvatures and concave or convex shapes).

For the cases of more than two random variables, the shortest distance from MPP, denoted as $z^* = (z_1^*, z_2^*, \ldots, z_n^*)$, to the origin is given by

$$\beta_{HL} = \sqrt{\sum_{i=1}^{n} z_i^{*2}}. \tag{2.20}$$

Again, if the limit state function is linear, i.e. $G(x) = a_0 + \sum_{i=1}^{n} a_i x_i$, and x_i are independent Gaussian random variables, $P_f = \Phi(-\beta_{HL})$ will be exact. In all other cases, P_f is an approximation and the problem can be solved by a constrained optimization algorithm to determine the minimum distance for a general nonlinear function, defined as

$$\beta_{HL} = \min\left(\sqrt{\sum_{i=1}^{n} z_i^2}\right) = \min(z^T z)^{1/2} \tag{2.21}$$

subject to $G(z) = 0$, equivalently $G(x) = 0$.

β_{HL} (and the design point) should be obtained through a computer iteration program satisfying the conditions given in equation (2.21).

Analytical techniques for finding the minimum distance developed by Shinozuka [108], and Ang and Tang [109], which provided the basis for performing the iteration process, are presented in Appendix II.

The Hasofer–Lind method [22] has the following notable properties:

1. **Invariance**, as the shortest distance from the origin to the limit surface in the reduced space, β_{HL}, is invariant. This is because, regardless of the form in which the limit state equation is written, its geometric shape and the distance from the origin remain the same. Also, the designer does not have to specify which variables are loads and which are resistances. Any change of the resistance definition consistent with the rules of mechanics will lead to the same failure region, and therefore the same safety index.
2. **Freedom of choice of basic variables.** The set of the basic variables need not have any physical significance as long as their joint distribution determines the distribution of the physical variables that enter into the problem.
3. **Capability of dealing with correlated basic variables, stress reversal and multiple failure modes.**[5]

The solution to the problem of invariance by β_{HL} represented significant progress in structural reliability theory [8]. Many enhanced applications of the FORM method can be found in the literature [110–114].

---

5. In the case of multiple failure modes, the safety index is simply the minimum of the safety indices determined by the various modes of failure.

It should be emphasized that, being a linear approximation method, there are two conditions for FORM to be accurate:

1. All the random variables are independent and Gaussian, and in the normal space.
2. The linear approximation to the failure surface must be reasonable.[6]

The approximation will be more accurate the more closely these conditions are obeyed.

2.4 RESPONSE SURFACE METHOD

To analyze the reliability of a very large, complex engineering structure with a number of random variables, the response surface method (RSM) can be used as an efficient tool to approximate the unknown limit state that may otherwise require costly numerical solutions.

Building an appropriate response surface model (which can be of a polynomial form, of first order or second order, or even combinations) and choosing a relevant sampling design method are the fundamental techniques of the RSM. To decide which of them to use, much depends on the nature of the problem.

2.4.1 Response Surface Models and Fitting Techniques

Four response surface models are usually selected and employed to replace the unknown limit state function. They are:

Type I model: $G(x) = y = \beta_0 + \sum_{j=1}^{k} \beta_j x_j + \varepsilon$

Type II model: $G(x) = y = \beta_0 + \sum_{j=1}^{k} \beta_j x_j + \sum_{i<j=2}^{k} \sum \beta_{ij} x_i x_j + \varepsilon$

Type III model: $G(x) = y = \beta_0 + \sum_{j=1}^{k} \beta_j x_j + \sum_{j=1}^{k} \beta_{jj} x_{jj}^2 + \varepsilon$

Type IV model: $G(x) = y = \beta_0 + \sum_{j=1}^{k} \beta_j x_j + \sum_{j=1}^{k} \beta_{jj} x_{jj}^2 + \sum_{i<j=2}^{k} \sum \beta_{ij} x_i x_j + \varepsilon.$

The type I model is a first-order polynomial, while type II is first order with interaction terms.[7] Type III is a second-order polynomial, as is type IV

---

6. It is more reasonable with a lower probability of failure (farther away from the origin), as discussed earlier.

7. This type model is treated as a second-order polynomial by some researchers. In the author's research, however, it will follow the terminology given in Myers and Montgomery's book [44].

but with interaction terms. The notations in the above equations will be explained in the following paragraphs.

In the reliability analysis of a structural problem, the observed response data are often obtained through a finite element program. It is supposed that a response y is influenced by k independent variables and n observations are obtained, i.e. n FE runs are performed. Suppose that a type I model is selected to describe the problem. We have

$$G(x) = y = \beta_0 + \sum_{j=1}^{k} \beta_j X + \varepsilon. \tag{2.22}$$

In mathematical terms, the above equation is a *multiple linear regression model* with k *regressor* variables. The parameters β_j ($j = 0, 1, \ldots, k$) are called the *regression coefficients* [44]. ε is the *error term*, which represents a statistical error. ε is assumed to have a normal distribution with zero mean and variance σ^2. An observation of the above equation is represented by

$$y_i = \beta_0 + \sum_{j=1}^{k} \beta_{ij} X_{ij} + \varepsilon_i. \tag{2.23}$$

Equation (2.22) can be written in matrix notation as

$$y = X\beta + \varepsilon, \tag{2.24}$$

where y is an $n \times 1$ vector of the observations (responses), X is an $n \times p$ matrix of the independent variables (where $p = k + 1$), β is a $p \times 1$ vector of the regression coefficients, and ε is an $n \times 1$ vector of random errors.

The *least squares estimation technique* is now applied. In order to minimize the following equation:

$$L = \sum_{i=1}^{n} \varepsilon_i^2 = \sum_{i=1}^{n} \left(y_i - \beta_0 - \sum_{j=1}^{k} \beta_{ij} X_{ij} \right)^2 = \varepsilon^T \varepsilon = (y - X\beta)^T (y - X\beta), \tag{2.25}$$

the least squares estimators, b, must satisfy

$$\frac{\partial L}{\partial \beta}\bigg|_b = -2X^T y + 2X^T X b = 0. \tag{2.26}$$

Then b can be solved via the matrix form:[8]

$$b = (X^T X)^{-1} X^T y. \tag{2.27}$$

The fitted model will be

$$\hat{y} = Xb, \tag{2.28}$$

---

8. If X is a square matrix, it simply has $b = X^{-1}y$.

and in scalar notation

$$\hat{y}_i = b_0 + \sum_{j=1}^{k} b_{ij} X_{ij}. \qquad (2.29)$$

2.4.2 Sampling Design Methods

How to select different sample points in the variable space is very important in the RSM applications, for it directly influences the efficiency and accuracy of the response surface model fitting. There are a number of sample point selection methods suitable for different types of RSM applications. Two of them are widely used, i.e. *Koshal design* and *central composite design* (CCD).

Koshal Design

Koshal design is a *saturated* design method that requires only the minimum runs for fitting a response surface model, for which the term "saturated" is given. The total number of sample points selected for fitting a first-order model (type I) is therefore $k + 1$, and for fitting a second-order model (type III) is $2k + 1$. The X matrix, in equation (2.24), for fitting a type I model with three coded variables[9] is of the form:

$$X = \begin{array}{c} \begin{array}{ccc} x_1 & x_2 & x_3 \end{array} \\ \begin{bmatrix} 1 & 0 & 0 & 0 \\ 1 & 1 & 0 & 0 \\ 1 & 0 & 1 & 0 \\ 1 & 0 & 0 & 1 \end{bmatrix} \end{array}$$

and for a type III model is

$$X = \begin{array}{c} \begin{array}{ccccccc} x_1 & x_2 & x_3 & x_1^2 & x_2^2 & x_3^2 \end{array} \\ \begin{vmatrix} 1 & 0 & 0 & 0 & 1 & 0 & 0 \\ 1 & 1 & 0 & 0 & 0 & 1 & 0 \\ 1 & 0 & 1 & 0 & 0 & 0 & 1 \\ 1 & 0 & 0 & 1 & 0 & 0 & 0 \\ 1 & 2 & 0 & 0 & 4 & 0 & 0 \\ 1 & 0 & 2 & 0 & 0 & 4 & 0 \\ 1 & 0 & 0 & 2 & 0 & 0 & 4 \end{vmatrix} \end{array}.$$

---

9. *Coded variable* is the terminology adopted in response surface methodology for standardized random variables.

Central Composite Design (CCD)

The central composite design sampling method is widely used in response surface applications. By selecting *corner*, *axial*, and *center* points, it is an ideal solution for fitting a second-order response surface model [44]. However, as it requires a relatively large number of sample points, the CCD method should only be chosen in a later stage of the RSM application when the total number of important variables is reduced to an acceptable figure.

For example, a type III second-order model is proposed for a two-random-variable response surface problem and the CCD method is chosen to select the sample points. As illustrated in Figure 2.4, in terms of the coded variables, the design will have four runs at the corners of the square $(-1, -1)$, $(1, -1)$, $(-1, 1)$, $(1,1)$; one run at the center point $(0,0)$; and another four axial runs at $(-\sqrt{2},0)$, $(\sqrt{2},0)$, $(0,-\sqrt{2})$, $(0,\sqrt{2})$. The total number of sample points selected for fitting such a type III model is 9 (determined by the equation $2^k + 2k + 1$),[10] while the minimum number of runs for fitting such model, in a saturated sampling method, is 5 (determined by the equation $2k + 1$). Thus when k is relatively large, the computational cost of running a finite element program using the CCD method is considerably higher.

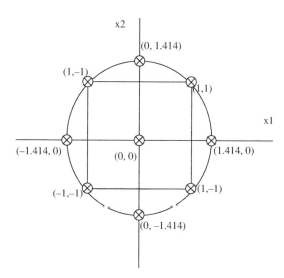

FIGURE 2.4   Central composite design with two variables.

---

10. CCD design normally requires more than one run at the center point, so that the total number of runs will often exceed $2^k + 2k + 1$ [44]. However, in the author's research, because the finite element program cannot generate different results for replicate runs, only one run at the center point is performed.

The variable X matrix for fitting such a type III model with two coded variables is

$$
X =
\begin{matrix}
& x_1 & x_2 & x_1^2 & x_2^2 \\
\begin{bmatrix}
1 & -1 & -1 & 1 & 1 \\
1 & 1 & -1 & 1 & 1 \\
1 & -1 & 1 & 1 & 1 \\
1 & 1 & 1 & 1 & 1 \\
1 & -1.414 & 0 & 2 & 0 \\
1 & 1.414 & 0 & 2 & 0 \\
1 & 0 & -1.414 & 0 & 2 \\
1 & 0 & 1.414 & 0 & 2 \\
1 & 0 & 0 & 0 & 0
\end{bmatrix}
\end{matrix}.
$$

The CCD method also maintains the rotatability of the variation, which is helpful in maintaining the accuracy of model fitting [44].

Intermediate between saturated design and CCD design, there is another popular design method called the *factorial design method*, which selects only the corner points. Table 2.1 summarizes the number of distinct sample points required by these three different design methods for fitting different RS models.

Summary

The type I RS model plus the saturated design method is the most efficient response surface fitting approach, in terms of the number of FE runs required.

TABLE 2.1 Sample Points Required by Different Design Methods for Different RS Models

RS Model Type	Saturated Design	$2^k/3^k$ Factorial Design	Central Composite Design (CCD) Design	Plus n_c Center Points
I	$k+1$	$2^k/3^k$	–	$+n_c$
II	$k + k(k-1)/2 + 1$	$2^k/3^k$	–	$+n_c$
III	$2k+1$	–	$2^k + 2k + 1$	$+n_c$
IV	$2k + k(k-1)/2 + 1 = (k+2)(k+1)/2$	–	$2^k + 2k + 1$	$+n_c$

This approach is adopted in fitting the covariance matrix of the stiffness matrix, which will be presented in Chapter 6.

2.5 PROBLEMS OF APPLYING FORM AND RSM METHODS TO DYNAMIC SYSTEMS

As discussed in Chapter 1, the current work aims to deliver an efficient reliability analysis method to dynamic systems using probabilistic information. Fundamentals of the modal analysis of dynamic systems, which are essential to the current research, are briefly reviewed and presented in Appendix III. As addressed earlier (Table 1.2), in addition to the obvious problem of high computational cost in running the direct Monte Carlo method, there are also difficulties in applying the FORM and RSM methods to dynamic systems.

2.5.1 Problematic Failure Surfaces for FORM Applications

As demonstrated previously, for the FORM method to work accurately, it requires that the failure surface is linear or close to linear in the space of random variables. This implies that the safety margin is a smooth and monotonic function. While this is often true for static systems, it is often not the case for dynamic systems.

This problem is illustrated in Figure 2.5 by a 2-DOF spring−mass dynamic system with two random mass variables (with mean values 25 and 25 kg respectively). The system settings and the safety margin equation are listed

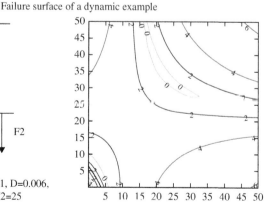

Failure surface of a dynamic example

$d_1 = ([K]) + i[D] - \omega^2[M])^{-1}\{F\}(1)$

Omega_f=7, eta = 0.05, F2=1, D=0.006, k1=400, k2=600, M1=25, M2=25

$g(m_1, m_2) = D - |d_1|$

$= D - |([K] + i[D] - \omega^2[M])^{-1}\{F\}(1)|$

Safety Margin contour plot over m1 and m2

FIGURE 2.5 Non-monotonic safety margins of a dynamic system.

in the figure. The two natural frequencies are 2.59 and 7.57 Hz, and the excitation frequency is 7 Hz.

Contours of the safety margins are plotted for the two random variables. Clearly, in contrast to the expected smooth and fairly linear failure surface depicted in Figure 2.3, the failure surface(s) shown in this example is multiply connected and non-monotonic, due to the resonance problem.

As demonstrated in Figure 2.5, the failure surface has minima at the two resonant frequencies. Obviously, this will be problematic for the FORM method to target the right failure surface in determining the safety index and the design point [103,104]. It is therefore difficult and risky to apply the FORM method to a dynamic system, as the result could be wrong.

2.5.2 Inaccuracy of RSM in Predicting the Dynamic Response

As discussed before, the RSM is used as a surrogate to the unknown limit state function performed by the FE method. However, it cannot predict the dynamic response accurately due to the complexity of the dynamic behavior [100,101].

Margetson [100,101] identified such a problem during the analysis of the forced vibration of a simply supported flat plate. The input random variables were the *force frequency, geometry and material properties*, and the fitted output was the *displacement* of a particular node. Two sets of data were used for the evaluation. One corresponded to the baseline dataset (mean values) and another to some alterations of the property variables. Two different results of the first mode analysis, compared to the finite element solutions, are plotted in Figure 2.6. It shows a significant discrepancy between the solid-line

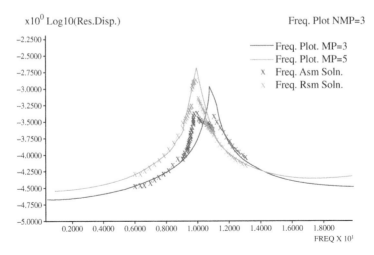

FIGURE 2.6 Comparison of the analytical solution and overlapping response surface solution for the first mode representation. *(Reproduced from Ref. [101])*

curves of the FE response solution and the star-line curves of the RSM approximation.

The investigation has concluded that the RSM fitting was inaccurate in predicting the dynamic response, and therefore cannot be used for further reliability analysis. The RSM method was eventually rejected in Margetson's project to directly predict the dynamic responses [100,101].

2.6 OPTIMIZATION SOLUTION THROUGH MODAL ANALYSIS

To overcome the above problems, i.e. the large volume FE runs of MCS, the problematic failure surface of FORM and the poor response prediction of RSM, the new proposed approach should make the following considerations:

1. To improve efficiency: The new approach should be able to work only on the mean eigen solution and derivations. This would mean that, instead of running a large number of inefficient FE calls (in solving either the inversions or eigen problems) like MCS, it just takes the mean values of the input variables and performs the eigen solver once to get the corresponding modal properties.
2. To construct a fairly linear failure surface: The new approach should seek an approximate model that defines some new modal parameters on which a fairly smooth and linear failure surface could be constructed, instead of the non-monotonic failure surface built upon the original random variables like FORM.
3. To employ the RSM in a better way: Rather than directly fitting over the dynamic response and eigen solution, RSM can be employed to efficiently build analytical models describing some other relationships in the reliability analysis process, which may otherwise need costly computation efforts, for example to obtain a required covariance matrix.

Based on these considerations, a new combined approach was proposed and investigated. It started with an approximate modal model (with a new set of defined parameters) being developed, based on the receptance elements and modal properties. Corresponding perturbation algorithms were then developed to map the statistical information from the original spatial parameters to the defined parameters in order to apply the FORM or MCS method.

The general modal expression of the receptance matrix is therefore essential to this new approach. The derivation of such an expression is discussed in Appendix III.

Equation (AIII.26) gives a general expression for the receptance matrix in terms of the modal properties. Further extraction of a single element in the response vector will provide a basis for developing an approximation model of dynamic response, on which the proposed perturbation algorithms are derived. This will be presented in the next chapter.

Chapter 3

Theoretical Fundamentals of the Perturbation Approach

This chapter presents the technical fundamentals of the core of the proposed combined approach, i.e. a perturbation approach, including the definition of the newly defined parameters and new safety margin, and the derivation of the first two moments of the new parameters.

3.1 DEFINITION OF THE NEW PARAMETERS AND SAFETY MARGIN

As explained in Appendix III, the response amplitude vector can be expressed in terms of the modal properties (equation AIII.26), i.e.

$$\{\overline{X}\} = \sum_{r=1}^{n} \frac{\{\phi\}_r^T \{\overline{F}\} \{\phi\}_r}{(\omega_r^2 - \omega^2) + i\eta_r \omega_r^2}. \tag{3.1}$$

Extracting one element from the response vector, denoted as $\overline{X}_j$, gives

$$\overline{X}_j = \sum_{r=1}^{n} \frac{\phi_{r,j} F_k \phi_{r,k}}{(\omega_r^2 - \omega^2) + i\eta_r \omega_r^2}. \tag{3.2}$$

Equation (3.2) represents the response at coordinate j due to a single harmonic force excitation applied at the coordinate k, and indicates that the vector $\{F\}$ has just one non-zero element F_k.[1]

Considering equation (AIII.15), a single element in the receptance FRF matrix is denoted as

$$\alpha_{jk}(\omega) = \left(\frac{\overline{X}_j}{F_k}\right) \quad F_m = 0; \quad m = 1, \dots, n; \quad m \neq k. \tag{3.3}$$

---

1. Please note that this is the case of a single force, so F_k is a real scalar and assumed to be deterministic in the current research. However, if multiple forces are involved, equation (3.2) will become complex. The corresponding analysis of multiple forces is presented in Appendix IV, though in this chapter and later examples, it is assumed that there is only a single force applied.

Reliability Analysis of Dynamic Systems.

By substituting equation (3.2) into (3.3), α_{jk} can be written in terms of the modal properties as

$$\alpha_{jk}(\omega) = \frac{\overline{X}_j}{F_k} = \sum_{r=1}^{n} \frac{\phi_{r,j}\phi_{r,k}}{(\omega_r^2 - \omega^2) + i\eta_r\omega_r^2} \tag{3.4}$$

or

$$\alpha_{jk}(\omega) = \sum_{r=1}^{n} \frac{\overline{A}_{jk,r}}{(\omega_r^2 - \omega^2) + i\eta_r\omega_r^2}, \tag{3.5}$$

where $\overline{A}_{jk,r}$ is the *modal constant*, a constant for given r, j, and k [77].

The receptance matrix $\alpha(\omega)$ possesses *symmetric* and *interrelationship* properties, i.e. $\alpha_{jk}(\omega) = \overline{X}_j/F_k = \alpha_{kj}(\omega) = \overline{X}_k/F_j$ and $\overline{A}_{jk,r} = \phi_{r,j}\phi_{r,k}$, $\overline{A}_{jj,r} = \phi_{r,j}^2$, $\overline{A}_{kk,r} = \phi_{r,k}^2$, which indicates that if one line (or column) of the matrix is determined, the whole matrix can be evaluated [77]. This is very useful to simplify the programming effort in assembling the relevant matrices.

It is the response vector and its element, e.g. $\overline{X}_j$, that is normally involved in constructing the required safety margin. Given that the *maximum allowed displacement* value at node j is denoted as $X_{j\mathrm{max}}$, a real positive number, the safety margin can be defined as

$$M = X_{j\mathrm{max}} - |\overline{X}_j|. \tag{3.6}$$

It should be noted that $\overline{X}_j$ in equations (3.2), (3.4), and (3.5) is complex.

The following expressions are technically equivalent to equation (3.6) in representing the reliability requirement:

$$M = X_{j\mathrm{max}}^2 - |\overline{X}_j|^2 \tag{3.7}$$

$$M = \ln X_{j\mathrm{max}}^2 - \ln |\overline{X}_j|^2. \tag{3.8}$$

They should yield the same results by the H-L FORM method. In order to maintain consistency with the response analysis (presented later for detailed examples), equation (3.8) is adopted as the safety margin equation in the current research.

To simplify the complex expression, the modulus of $\overline{X}_j$ is taken, which can be determined by its complex conjugate of the summation, i.e.

$$|\overline{X}_j|^2 = \left(\sum_{r=1}^{n} \frac{\phi_{r,j}F_k\phi_{r,k}}{(\omega_r^2 - \omega^2) + i\eta_r\omega_r^2} \right) \left(\sum_{p=1}^{n} \frac{(\phi_{p,j}F_k\phi_{p,k})}{(\omega_p^2 - \omega^2) + i\eta_p\omega_p^2} \right)^*. \tag{3.9}$$

It should be noted that, in the proportionally damped or undamped cases, $\phi_{r,j}$, $\phi_{r,k}$ are all real [77]; however, F_k can be complex. Thus, the above equation becomes

$$|\overline{X}_j|^2 = \left(\sum_{r=1}^{n} \frac{\phi_{r,j}F_k\phi_{r,k}}{(\omega_r^2 - \omega^2) + i\eta_r\omega_r^2} \right) \left(\sum_{p=1}^{n} \frac{(\phi_{p,j}F_k\phi_{p,k})^*}{(\omega_p^2 - \omega^2) - i\eta_p\omega_p^2} \right), \qquad (3.10)$$

or in an expanded form:

$$|\overline{X}_j|^2 = \underbrace{\frac{\phi_{1,j}F_k\phi_{1,k}}{(\omega_1^2 - \omega^2) + i\eta_1\omega_1^2}}_{b_1 \quad r=1} \overbrace{\left(\underbrace{\frac{(\phi_{1,j}F_k\phi_{1,k})^*}{(\omega_1^2 - \omega^2) - i\eta_1\omega_1^2}}_{c_1 \quad p=1} + \ldots + \frac{(\phi_{n,j}F_k\phi_{n,k})^*}{(\omega_n^2 - \omega^2) - i\eta_n\omega_n^2} \right)}^{T}$$

$$+ \ldots$$

$$+ \underbrace{\frac{\phi_{n,j}F_k\phi_{n,k}}{(\omega_n^2 - \omega^2) + i\eta_n\omega_n^2}}_{b_n \quad r=n} \overbrace{\left(\underbrace{\frac{(\phi_{1,j}F_k\phi_{1,k})^*}{(\omega_1^2 - \omega^2) - i\eta_1\omega_1^2}}_{} + \ldots + \frac{(\phi_{n,j}F_k\phi_{n,k})^*}{(\omega_n^2 - \omega^2) - i\eta_n\omega_n^2} \right)}^{T} \Bigg\}^{n.}$$

$$\underbrace{\hphantom{XX}}_{c_n \quad p=n}$$

$$(3.11)$$

Each row on the r.h.s. of equation (3.11) involves a response component (term b) multiplied by each component (term c) in a summed response (term T). The physical meaning of such a summed product can be represented as in Figure 3.1. When the modal overlap factor[2] is low, the peaks of responses are well separated, as shown in the figure, which is often the case

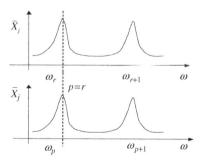

FIGURE 3.1   The two dominant terms in the modal summation at resonance.

---

2. *Modal overlap factor* is defined as $M = \omega\eta n$, where n is the modal density (average number of modes per unit frequency) [115]. As discussed in Chapter 1, FEA is capable of dealing with low modal overlap problems, while SEA normally requires high modal overlap.

for low-frequency dynamic systems. Considering resonance cases (which are of most concern), when $p = r$, b and c are both large. When $p \neq r$, at least one of b and c is small. The result of each row in equation (3.11) is therefore dominated by the product of two terms with the same natural frequencies (i.e. when $p = r$), such as b_1 and c_1, b_n and c_n respectively.

Thus, the modulus of $\overline{X}_j$ can be further approximated to

$$
|\overline{X}_j|^2 \approx \left.
\begin{array}{c}
\underbrace{\dfrac{\phi_{1,j}F_k\phi_{1,k}}{(\omega_1^2 - \omega^2) + i\eta_1\omega_1^2}}_{r=1} \underbrace{\dfrac{(\phi_{1,j}F_k\phi_{1,k})^*}{(\omega_1^2 - \omega^2) - i\eta_1\omega_1^2}}_{p=1} \\[1.0em]
+ \ldots \\[1.0em]
+ \underbrace{\dfrac{\phi_{n,j}F_k\phi_{n,k}}{(\omega_n^2 - \omega^2) + i\eta_n\omega_n^2}}_{r=n} \underbrace{\dfrac{(\phi_{n,j}F_k\phi_{n,k})^*}{(\omega_n^2 - \omega^2) - i\eta_n\omega_n^2}}_{p=n}
\end{array}
\right\} n
\tag{3.12}
$$

or alternatively in a *real* summation form as

$$
|\overline{X}_j|^2 \approx \sum_{r=1}^{n} \frac{|\phi_{r,j}F_k\phi_{r,k}|^2}{(\omega_r^2 - \omega^2)^2 + (\eta_r\omega_r^2)^2}.
\tag{3.13}
$$

Now, it is appropriate to define two *new parameters*, the *denominator* of equation (3.13), defined as d_r,

$$
d_r = (\omega_r^2 - \omega^2)^2 + (\eta_r\omega_r^2)^2,
\tag{3.14}
$$

and the *numerator* of the equation, defined as $r_{jk,r}$,

$$
r_{jk,r} = \phi_{r,j}F_k\phi_{r,k}.
\tag{3.15}
$$

Both parameters are *scalars* and constituted by modal parameters specific to r, j, and k.

Substituting the new parameters into equation (3.13) leads to

$$
|\overline{X}_j|^2 = \sum_{r=1}^{n} \frac{|r_{jk,r}|^2}{d_r}.
\tag{3.16}
$$

Equation (3.16) provides an approximation model of the system response.

The *safety margin* can now be presented in terms of the defined parameters $r_{jk,r}$ and d_r. Substituting equation (3.16) into equation (3.8) gives

$$
M = \ln X_{jmax}^2 - \ln|\overline{X}_j|^2 = 2\ln X_{max} - \ln\left(\sum_{r=1}^{n} \frac{|r_{jk,r}|^2}{d_r}\right).
\tag{3.17}
$$

Again, considering resonance cases, i.e. when the excitation frequency ω is equal to one of the system natural frequencies ω_r, d_r (3.14) would reach its minimum for that frequency. Equation (3.16) will then be dominated by the resonance mode, i.e. $\left|\overline{X}_j\right|^2 \approx \left|r_{jk,r}\right|^2 / d_r$ (or, as used in later studies, $\left|\overline{X}_j\right|^2 = \left|r_{jk,r}\right|^2 / d_r + \Sigma_{i \neq r}^n \left|r_{jk,i}\right|^2 / d_i$, where the term $\Sigma_{i \neq r}^n \left|r_{jk,i}\right|^2 / d_i$ is a constant value and it is small compared to the resonance mode term $\left|r_{jk,r}\right|^2 / d_r$). According to equation (3.17), in resonance cases, the new safety margin can then be further approximated by a linear combination of logarithm functions of the defined parameters d_r and $r_{jk,r}$, because the term $\ln\left|\overline{X}_j\right|^2$ can be further expanded into

$$\ln\left|\overline{X}_j\right|^2 \approx \ln\left(\frac{\left|r_{jk,r}\right|^2}{d_r}\right) = 2\ln\left(\left|r_{jk,r}\right|\right) - \ln(d_r). \qquad (3.18)$$

Fulfilling the (linearity) requirement (2) listed in Section 2.6, this approximation at resonance is an important outcome of the introduction of the new parameters.

3.2 DERIVATION OF THE TWO MOMENTS OF THE NEW PARAMETERS

In order to apply FORM on the defined parameters, the first two moments, i.e. the mean and variance, or the corresponding *covariance matrices*, of d_r and $r_{jk,r}$, must be known. It is assumed that the first two moments of any random variables involved are available, including the covariance matrices of the whole stiffness matrix $[K]$ and mass matrix $[M]$.[3] Therefore, the task was to derive the required moments based on the known information.

As illustrated in Figure 3.2, the task consists of two subtasks. Firstly, the covariance matrices of the defined parameters, denoted as COV in the figure,

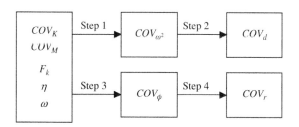

FIGURE 3.2   Derivation steps of the two moments of the defined parameters.

---

3. This assumes the statistical information of all the random geometry and property variables is known, and the covariance matrices of the mass matrix and the stiffness matrix, COV_M, COV_K, can be obtained. How to derive COV_M, COV_K from the original variables will be presented in Chapter 6.

were developed. Perturbation techniques were employed with respect to the alteration of spatial parameters. Four steps were conducted to derive the covariance matrices of ω_j^2, then d_r and ϕ_j, then $r_{jk,r}$ respectively. These steps were based on the known statistical information including the covariance matrices of $[K]$ and $[M]$, and other deterministic parameters such as F_k, η, ω. Secondly, the algebraic equations determining the mean values of the defined parameters were developed. Details of the derivation are reported in the following sections.

3.2.1 Derivation of the Covariance Matrix of the Modal Parameter ω^2

The fundamental eigen equation gives [77]

$$\omega_j^2 M \phi_j = K \phi_j. \tag{3.19}$$

where ϕ_j is the jth column of the mass-normalized modal matrix, defined in equation (AIII.12).

Assuming that small changes are applied to the original spatial parameter matrices K and M *around their mean values*:[4]

$$\begin{aligned} M &\to M + \Delta M \\ K &\to K + \Delta K. \end{aligned} \tag{3.20}$$

The corresponding modal parameters would respond with changes as

$$\begin{aligned} \omega_j^2 &\to \omega_j^2 + \Delta \omega_j^2 \\ \phi_j &\to \phi_j + \Delta \phi_j. \end{aligned} \tag{3.21}$$

Substituting (3.21) and (3.20) into equation (3.19) gives

$$(\omega_j^2 + \Delta \omega_j^2)(M + \Delta M)(\phi_j + \Delta \phi_j) = (K + \Delta K)(\phi_j + \Delta \phi_j). \tag{3.22}$$

Expanding the above equation yields

$$\begin{aligned} &\omega_j^2 M \phi_j + \omega_j^2 M \Delta \phi_j + \omega_j^2 \Delta M \phi_j + \omega_j^2 \Delta M \Delta \phi_j \\ &+ \Delta \omega_j^2 M \phi_j + \Delta \omega_j^2 M \Delta \phi_j + \Delta \omega_j^2 \Delta M \phi_j + \Delta \omega_j^2 \Delta M \Delta \phi_j \\ &= K \phi_j + K \Delta \phi_j + \Delta K \phi_j + \Delta K \Delta \phi_j. \end{aligned} \tag{3.23}$$

Ignoring the second-order Δ terms (because that they are much smaller changes) in equation (3.23) leads to

$$\omega_j^2 M \phi_j + \omega_j^2 M \Delta \phi_j + \omega_j^2 \Delta M \phi_j + \Delta \omega_j^2 M \phi_j \approx K \phi_j + K \Delta \phi_j + \Delta K \phi_j. \tag{3.24}$$

---

4. Unless otherwise stated, all notations of random variables hereafter in this chapter represent their mean values.

Because $\omega_j^2 M\phi_j = K\phi_j$, the above equation is further simplified to

$$\omega_j^2 M\Delta\phi_j + \omega_j^2 \Delta M\phi_j + \Delta\omega_j^2 M\phi_j \approx K\Delta\phi_j + \Delta K\phi_j. \tag{3.25}$$

Pre-multiplying by ϕ_j^T on both sides of the above equation gives

$$\phi_j^T \omega_j^2 M\Delta\phi_j + \phi_j^T \omega_j^2 \Delta M\phi_j + \phi_j^T \Delta\omega_j^2 M\phi_j \approx \phi_j^T K\Delta\phi_j + \phi_j^T \Delta K\phi_j. \tag{3.26}$$

Because $\phi_j^T \omega_j^2 M\Delta\phi_j = \phi_j^T K\Delta\phi_j$, the following equation is obtained:

$$\phi_j^T M\phi_j \Delta\omega_j^2 \approx \phi_j^T \Delta K\phi_j - \phi_j^T \Delta M\phi_j \omega_j^2. \tag{3.27}$$

Because $\phi_j^T M\phi_j = 1$, equation (3.27) finally becomes [78]

$$\Delta\omega_j^2 \approx \phi_j^T \Delta K\phi_j - \phi_j^T \Delta M\phi_j \omega_j^2. \tag{3.28}$$

The above equation will be used to develop the covariance matrix of the modal parameters ω_j^2.

The full matrices of ΔK and ΔM are written as

$$\Delta K = \begin{pmatrix} \Delta k_{11} & \Delta k_{21} & \ldots & \Delta k_{1n} \\ \Delta k_{21} & \Delta k_{22} & \ldots & \Delta k_{2n} \\ \ldots & \ldots & \ldots & \ldots \\ \Delta k_{n1} & \Delta k_{n2} & \ldots & \Delta k_{nn} \end{pmatrix}_{n \times n} \tag{3.29}$$

$$\Delta M = \begin{pmatrix} \Delta m_{11} & \Delta m_{21} & \ldots & \Delta m_{1n} \\ \Delta m_{21} & \Delta m_{22} & \ldots & \Delta m_{2n} \\ \ldots & \ldots & \ldots & \ldots \\ \Delta m_{n1} & \Delta m_{n2} & \ldots & \Delta m_{nn} \end{pmatrix}_{n \times n}. \tag{3.30}$$

Rearranging ΔK and ΔM into $n \times 1$ vectors (column by column), denoted as k and m respectively, gives

$$k = \begin{pmatrix} \Delta k_{11} \\ \Delta k_{21} \\ \ldots \\ \Delta k_{n1} \\ \Delta k_{12} \\ \Delta k_{22} \\ \ldots \\ \Delta k_{n2} \\ \ldots \\ \Delta k_{nn} \end{pmatrix}_{n^2 \times 1} \tag{3.31}$$

and

$$m = \begin{pmatrix} \Delta m_{11} \\ \Delta m_{21} \\ \dots \\ \Delta m_{n1} \\ \Delta m_{12} \\ \Delta m_{22} \\ \dots \\ \Delta m_{n2} \\ \dots \\ \Delta m_{nn} \end{pmatrix}_{n^2 \times 1} . \tag{3.32}$$

The vector $\Delta \omega_j^2$ (equation (3.28)) can be expressed in terms of k, m, and ϕ_j, ϕ_j^T:

$$\Delta \omega^2 = \begin{pmatrix} \Delta \omega_1^2 \\ \Delta \omega_2^2 \\ \dots \\ \Delta \omega_n^2 \end{pmatrix}_{n \times 1}$$

$$= \begin{pmatrix} \phi_{1,1}\phi_{1,1} & \phi_{1,2}\phi_{1,1} & \cdots & \phi_{1,n}\phi_{1,1} & \cdots & \phi_{1,n}\phi_{1,n} \\ \phi_{2,1}\phi_{2,1} & \phi_{2,2}\phi_{2,1} & & \phi_{2,n}\phi_{2,1} & & \phi_{2,n}\phi_{2,n} \\ \dots & & & & & \dots \\ \dots & & & \phi_{j,r}\phi_{j,s} & & \dots \\ \dots & & & & & \dots \\ \phi_{n,1}\phi_{n,1} & \phi_{n,2}\phi_{n,1} & & \phi_{n,n}\phi_{n,1} & & \phi_{n,n}\phi_{n,n} \end{pmatrix}_{n \times n^2} \begin{pmatrix} \Delta k_{11} \\ \Delta k_{21} \\ \dots \\ \Delta k_{n1} \\ \dots \\ \Delta k_{rs} \\ \dots \\ \Delta k_{nn} \end{pmatrix}_{n^2 \times 1}$$

$$- \begin{pmatrix} \omega_1^2\phi_{1,1}\phi_{1,1} & \omega_1^2\phi_{1,2}\phi_{1,1} & \cdots & \omega_1^2\phi_{1,n}\phi_{1,1} & \cdots & \omega_1^2\phi_{1,n}\phi_{1,n} \\ \omega_2^2\phi_{2,1}\phi_{2,1} & \omega_2^2\phi_{2,2}\phi_{2,1} & & \omega_2^2\phi_{2,n}\phi_{2,1} & & \omega_2^2\phi_{2,n}\phi_{2,n} \\ \dots & & & & & \dots \\ \dots & & & \omega_j^2\phi_{j,r}\phi_{j,s} & & \dots \\ \dots & & & & & \dots \\ \omega_n^2\phi_{n,1}\phi_{n,1} & \omega_n^2\phi_{n,2}\phi_{n,1} & & \omega_n^2\phi_{n,n}\phi_{n,1} & & \omega_n^2\phi_{n,n}\phi_{n,n} \end{pmatrix}_{n \times n^2} \begin{pmatrix} \Delta m_{11} \\ \Delta m_{21} \\ \dots \\ \Delta m_{n1} \\ \dots \\ \Delta m_{rs} \\ \dots \\ \Delta m_{nn} \end{pmatrix}_{n^2 \times 1} .$$

$$\tag{3.33}$$

It should be noted that, although the dimensions of the matrices in the above equation reach n^2, the matrices can be greatly truncated in practice if only a small number of mass or stiffness elements are random (which is often the case in many engineering problems [82]). The mode shape matrices involved can correspondingly be "shrunk" and the efficiency of the calculation process will be greatly improved.

Now, defining the relevant terms as

$$\Omega = \Delta\omega^2 = \begin{pmatrix} \Delta\omega_1^2 \\ \Delta\omega_2^2 \\ \cdots \\ \Delta\omega_n^2 \end{pmatrix}_{n \times 1} \qquad (3.34)$$

$$A = \begin{pmatrix} \phi_{1,1}\phi_{1,1} & \phi_{1,2}\phi_{1,1} & \cdots & \phi_{1,n}\phi_{1,1} & \cdots & \phi_{1,n}\phi_{1,n} \\ \phi_{2,1}\phi_{2,1} & \phi_{2,2}\phi_{2,1} & & \phi_{2,n}\phi_{2,1} & & \phi_{2,n}\phi_{2,n} \\ \cdots & & & & & \\ \cdots & & & \phi_{j,r}\phi_{j,s} & \cdots & \\ \cdots & & & & & \cdots \\ \phi_{n,1}\phi_{n,1} & \phi_{n,2}\phi_{n,1} & & \phi_{n,n}\phi_{n,1} & & \phi_{n,n}\phi_{n,n} \end{pmatrix}_{n \times n^2}, \qquad (3.35)$$

and

$$B = \begin{pmatrix} \omega_1^2\phi_{1,1}\phi_{1,1} & \omega_1^2\phi_{1,2}\phi_{1,1} & \cdots & \omega_1^2\phi_{1,n}\phi_{1,1} & \cdots & \omega_1^2\phi_{1,n}\phi_{1,n} \\ \omega_2^2\phi_{2,1}\phi_{2,1} & \omega_2^2\phi_{2,2}\phi_{2,1} & & \omega_2^2\phi_{2,n}\phi_{2,1} & & \omega_2^2\phi_{2,n}\phi_{2,n} \\ \cdots & & & & & \\ \cdots & & & \omega_j^2\phi_{j,r}\phi_{j,s} & \cdots & \\ \cdots & & & & & \cdots \\ \omega_n^2\phi_{n,1}\phi_{n,1} & \omega_n^2\phi_{n,2}\phi_{n,1} & & \omega_n^2\phi_{n,n}\phi_{n,1} & & \omega_n^2\phi_{n,n}\phi_{n,n} \end{pmatrix}_{n \times n^2}, \qquad$$

$$(3.36)$$

and substituting them into equation (3.33), the alternative equation for $\Delta\omega_j^2$ can be described in matrix form as

$$\Omega = Ak - Bm. \qquad (3.37)$$

The covariance matrix of Ω is defined as

$$C_\Omega = E[\Omega\Omega^T], \qquad (3.38)$$

where E represents "the expected value of", i.e. the mean value.

Substituting equation (3.37) into equation (3.38) and applying the relevant algebra and statistics techniques yields

$$
\begin{aligned}
C_\Omega &= E[(Ak - Bm)(Ak - Bm)^T] \\
&= E[(Ak - Bm)(k^T A^T - m^T B^T)] \\
&= E[Akk^T A^T - Akm^T B^T - Bmk^T A^T + Bmm^T B^T] \\
&= AE[kk^T]A^T - AE[km^T]B^T - BE[mk^T]A^T + BE[mm^T]B^T.
\end{aligned}
\tag{3.39}
$$

It is assumed that *k and m are independent of each other* when responding to small changes.[5] Therefore, equation (3.39) becomes

$$
C_\Omega = AE[kk^T]A^T - AE[k]E[m^T]B^T - BE[m]E[k^T]A^T + BE[mm^T]B^T. \tag{3.40}
$$

Because the small changes applied are around their mean values, the expected values of these changes are zero, i.e. $E[k] = E[m] = 0$. Thus, the above equation finally becomes

$$
C_\Omega = AC_k A^T + BC_m B^T, \tag{3.41}
$$

where C_k and C_m are the covariance matrices of k and m respectively, which are assumed to be known (refer to Chapter 6 for more details about how they are determined from the original random variables).

Thus, so far, given that the covariance matrices of the physical parameters C_k and C_m are available, the covariance matrix of the natural frequency squares C_Ω can be evaluated.

3.2.2 Derivation of the Covariance Matrix of the Defined Parameter d_r

The d_r definition equation (3.14) can be expanded as

$$
d_r = \omega_r^4 - 2\omega_r^2 \omega^2 + \omega^4 + \eta_r^2 \omega_r^4. \tag{3.42}
$$

Similarly, assuming that small changes are applied to ω_r^2 and consequently to d_r as

$$
\begin{aligned}
\omega_r^2 &\to \omega_r^2 + \Delta\omega_r^2 \\
d_r &\to d_r + \Delta d_r,
\end{aligned}
\tag{3.43}
$$

---

5. In real cases, the original random variables are sometimes correlated, which would affect the statistical properties of the mass and stiffness matrices. However, to reduce the complexity, it can be assumed that such effects are not statistically large and the effects can be accommodated in the derivation process of the covariance matrices of the mass and stiffness matrices, which will be presented in Chapter 6.

substituting them into the definition equation (3.14) gives

$$d_r + \Delta d_r = [(\omega_r^2 + \Delta\omega_r^2) - \omega^2]^2 + [\eta_r(\omega_r^2 + \Delta\omega_r^2)]^2. \tag{3.44}$$

Expanding the above equation and substituting equation (3.42) into the left part of equation (3.44) yields

$$\omega_r^4 - 2\omega_r^2\omega^2 + \omega^4 + \eta_r^2\omega_r^4 + \Delta d_r =$$
$$\omega_r^4 + 2\omega_r^2\Delta\omega_r^2 + \Delta\omega_r^4 - 2\omega_r^2\omega^2 - 2\Delta\omega_r^2\omega^2 + \omega^4 + \eta_r^2\omega_r^4 + 2\eta_r^2\omega_r^2\Delta\omega_r^2 + \eta_r^2\Delta\omega_r^4. \tag{3.45}$$

Canceling the equivalent terms on both sides of (3.45) yields

$$\Delta d_r = 2\omega_r^2\Delta\omega_r^2 + \Delta\omega_r^4 - 2\Delta\omega_r^2\omega^2 + 2\eta_r^2\omega_r^2\Delta\omega_r^2 + \eta_r^2\Delta\omega_r^4$$
$$= 2(1 + \eta_r^2)\omega_r^2\Delta\omega_r^2 + (1 + \eta_r^2)\Delta\omega_r^4 - 2\Delta\omega_r^2\omega^2. \tag{3.46}$$

Due to the fact that only small changes are applied, *ignoring the higher-order term* $\Delta\omega_r^4$ in equation (3.46) finally leads to

$$\Delta d_r \approx 2(\omega_r^2 - \omega^2 + \eta_r^2\omega_r^2)\Delta\omega_r^2. \tag{3.47}$$

Further, the whole vector of the small alternations of the defined variable d can be expressed as

$$\begin{pmatrix} \Delta d_1 \\ \Delta d_2 \\ \cdots \\ \Delta d_n \end{pmatrix}_{n\times 1} = 2 \begin{pmatrix} \omega_1^2 - \omega^2 + \eta_1^2\omega_1^2 & 0 & \cdots & 0 \\ 0 & \omega_2^2 - \omega^2 + \eta_2^2\omega_2^2 & \cdots & \cdots \\ \cdots & \cdots & \cdots & \cdots \\ \cdots & \cdots & \cdots & \omega_n^2 - \omega^2 + \eta_n^2\omega_n^2 \end{pmatrix}_{n\times n} \begin{pmatrix} \Delta\omega_1^2 \\ \Delta\omega_2^2 \\ \cdots \\ \Delta\omega_n^2 \end{pmatrix}_{n\times 1}. \tag{3.48}$$

Defining the relevant terms in the above equation as

$$D = \begin{pmatrix} \Delta d_1 \\ \Delta d_2 \\ \cdots \\ \Delta d_n \end{pmatrix}_{n\times 1} \tag{3.49}$$

and

$$H = 2 \begin{pmatrix} \omega_1^2 - \omega^2 + \eta_1^2\omega_1^2 & 0 & \cdots & 0 \\ 0 & \omega_2^2 - \omega^2 + \eta_2^2\omega_2^2 & \cdots & \cdots \\ \cdots & \cdots & \cdots & \cdots \\ \cdots & \cdots & \cdots & \omega_n^2 - \omega^2 + \eta_n^2\omega_n^2 \end{pmatrix}_{n\times n}, \tag{3.50}$$

and substituting them into equation (3.48), the alternative equation for the whole vector d can now be rewritten in matrix form as

$$D = H\Omega. \tag{3.51}$$

The covariance matrix of D is defined as

$$C_D = E[DD^T]. \tag{3.52}$$

Substituting equation (3.51) into equation (3.52) and applying the relevant techniques yields

$$
\begin{aligned}
C_D = E[DD^T] &= E[(H\Omega)(H\Omega)^T] \\
&= E[H\Omega\Omega^T H^T] = HE[\Omega\Omega^T]H^T \\
&= HC_\Omega H^T.
\end{aligned} \tag{3.53}
$$

Substituting equation (3.41) into equation (3.53) finally gives

$$C_D = H(AC_kA^T + BC_mB^T)H^T, \tag{3.54}$$

which indicates that the covariance matrix of D can be evaluated if the covariance matrices of the spatial parameters C_k and C_m are known.

3.2.3 Derivation of the Covariance Matrix of the Modal Parameter [Φ]

Equation (3.25), after the higher-order terms are ignored, can be rewritten as

$$K\Delta\phi_j + \Delta K\phi_j - (\omega_j^2 M\Delta\phi_j + \omega_j^2 \Delta M\phi_j + \Delta\omega_j^2 M\phi_j) \approx 0. \tag{3.55}$$

Pre-multiplying by ϕ_k^T on both sides yields

$$\phi_k^T K\Delta\phi_j + \phi_k^T \Delta K\phi_j - (\omega_j^2 \phi_k^T M\Delta\phi_j + \omega_j^2 \phi_k^T \Delta M\phi_j + \Delta\omega_j^2 \phi_k^T M\phi_j) \approx 0. \tag{3.56}$$

Equation (3.56) can be further rearranged as

$$\phi_k^T[-\omega_j^2 \Delta M + \Delta K]\phi_j - \Delta\omega_j^2 \delta_{kj} - \omega_j^2 \phi_k^T M\Delta\phi_j + \phi_k^T K\Delta\phi_j \approx 0, \tag{3.57}$$

where δ_{kj} is the *Kronecker delta*, which is defined as

$$\delta_{kj} = \begin{cases} 1 & k = j \\ 0 & k \neq j \end{cases}.$$

To solve the above equation for $\Delta\phi_j$, a *participation factor* is introduced. It assumes that the change of any mode shape w.r.t. any spatial changes can be expressed as a linear combination of all mode shapes that are independent of each other, i.e.

$$\Delta\phi_j = \sum_{r=1}^{n} a_r\phi_r, \tag{3.58}$$

where a_r is the *participation factor* for the rth mode.

Substituting equation (3.58) into equation (3.57) yields

$$\phi_k^T[-\omega_j^2\Delta M + \Delta K]\phi_j - \Delta\omega_j^2\delta_{kj} - \omega_j^2\phi_k^T M\left(\sum_{r=1}^{n} a_r\phi_r\right) + \phi_k^T K\left(\sum_{r=1}^{n} a_r\phi_r\right) \approx 0. \tag{3.59}$$

The above equation can be rearranged as

$$\phi_k^T[-\omega_j^2\Delta M + \Delta K]\phi_j - \Delta\omega_j^2\delta_{kj} - \omega_j^2\sum_{r=1}^{n} a_r\phi_k^T M\phi_r + \sum_{r=1}^{n} a_r\phi_k^T K\phi_r \approx 0. \tag{3.60}$$

By applying the orthogonality properties of the eigenvectors, the last two terms in the above equation can be simplified so that only the $r = k$ terms remain, i.e.

$$\phi_k^T[-\omega_j^2\Delta M + \Delta K]\phi_j - \Delta\omega_j^2\delta_{kj} - \omega_j^2 a_k + \omega_k^2 a_k \approx 0. \tag{3.61}$$

When $k = j$, the last two terms involving a_k will cancel, and equation (3.28) is obtained.

When $k \neq j$, solving for a_k according to equation (3.61) results in

$$a_k = \frac{\phi_k^T[-\omega_j^2\Delta M + \Delta K]\phi_j}{\omega_j^2 - \omega_k^2}. \tag{3.62}$$

By substituting equation (3.62) into equation (3.58) and adding the missing jth term (as a_k is only valid when $k \neq j$), $\Delta\phi_j$ is now expressed as

$$\Delta\phi_j = \sum_{k=1,k\neq j}^{n} a_k\phi_k + a_j\phi_j. \tag{3.63}$$

According to the mass orthogonality property (AIII.11), after the changes are applied, it should still satisfy

$$(\phi_j + \Delta\phi_j)^T(M + \Delta M)(\phi_j + \Delta\phi_j) = 1. \tag{3.64}$$

Expanding equation (3.64), applying the relevant orthogonality properties, and ignoring terms involving the two Δ terms finally yields an approximation equation:

$$\phi_j^T M\Delta\phi_j + \phi_j^T \Delta M\phi_j + \Delta\phi_j^T M\phi_j \approx 0. \tag{3.65}$$

Substituting equation (3.63) into equation (3.65) and expanding it yields

$$a_k \sum_{k=1,k\neq j}^{n} \phi_j^T M \phi_k + a_j \phi_j^T M \phi_j + \phi_j^T \Delta M \phi_j + a_k \sum_{k=1,k\neq j}^{n} \phi_k^T M \phi_j + a_j \phi_j^T M \phi_j = 0.$$

(3.66)

Applying the orthogonality properties to equation (3.66) finally gives

$$a_j = -\frac{1}{2}(\phi_j^T \Delta M \phi_j).$$

(3.67)

Substituting for a_k (3.62) and a_j (3.67) in equation (3.63), $\Delta\phi_j$ is finally expressed as

$$\Delta\phi_j = \sum_{k\neq j} \left[\frac{\phi_k^T[-\omega_j^2 \Delta M + \Delta K]\phi_j}{\omega_j^2 - \omega_k^2} \right] \phi_k - \frac{1}{2}(\phi_j^T \Delta M \phi_j)\phi_j.$$

(3.68)

By expanding the above equation and letting $\alpha_{jk} = \omega_j^2/(\omega_j^2 - \omega_k^2)$ and $\beta_{jk} = 1/(\omega_j^2 - \omega_k^2)$, $\Delta\phi_j$ can be described in full matrix form as

$$\Delta\phi_j = \begin{pmatrix} \Delta\phi_{j,1} \\ \Delta\phi_{j,2} \\ \cdots \\ \Delta\phi_{j,n} \end{pmatrix}_{n\times 1}$$

$$= \begin{pmatrix} \sum_{k\neq j}^{n}\beta_{jk}\phi_{k,1}\phi_{j,1}\phi_{k,1} & \sum_{k\neq j}^{n}\beta_{jk}\phi_{k,2}\phi_{j,1}\phi_{k,1} & \cdots & \sum_{k\neq j}^{n}\beta_{jk}\phi_{k,n}\phi_{j,1}\phi_{k,1} & \cdots & \sum_{k\neq j}^{n}\beta_{jk}\phi_{k,r}\phi_{j,s}\phi_{k,1} & \cdots & \sum_{k\neq j}^{n}\beta_{jk}\phi_{k,n}\phi_{j,n}\phi_{k,1} \\ \sum_{k\neq j}^{n}\beta_{jk}\phi_{k,1}\phi_{j,1}\phi_{k,2} & & & & & & & \\ \cdots & & & & & & & \\ & & & & \sum_{k\neq j}^{n}\beta_{jk}\phi_{k,r}\phi_{j,s}\phi_{k,t} & & \cdots & \\ \cdots & & & & & & & \\ \sum_{k\neq j}^{n}\beta_{jk}\phi_{k,1}\phi_{j,1}\phi_{k,n} & & & & & & \sum_{k\neq j}^{n}\beta_{jk}\phi_{k,n}\phi_{j,n}\phi_{k,n} \end{pmatrix}_{n\times n^2} \begin{pmatrix} \Delta k_{11} \\ \Delta k_{21} \\ \cdots \\ \Delta k_{n1} \\ \cdots \\ \Delta k_{rs} \\ \cdots \\ \Delta k_{nn} \end{pmatrix}_{n^2\times 1}$$

$$- \begin{pmatrix} \sum_{k\neq j}^{n}\alpha_{jk}\phi_{k,1}\phi_{j,1}\phi_{k,1} & \sum_{k\neq j}^{n}\alpha_{jk}\phi_{k,2}\phi_{j,1}\phi_{k,1} & \cdots & \sum_{k\neq j}^{n}\alpha_{jk}\phi_{k,n}\phi_{j,1}\phi_{k,1} & \cdots & \sum_{k\neq j}^{n}\alpha_{jk}\phi_{k,r}\phi_{j,s}\phi_{k,1} & \cdots & \sum_{k\neq j}^{n}\alpha_{jk}\phi_{k,n}\phi_{j,n}\phi_{k,1} \\ \sum_{k\neq j}^{n}\alpha_{jk}\phi_{k,1}\phi_{j,1}\phi_{k,2} & & & & & & & \\ \cdots & & & & & & & \\ & & & & \sum_{k\neq j}^{n}\alpha_{jk}\phi_{k,r}\phi_{j,s}\phi_{k,t} & & \cdots & \\ \cdots & & & & & & & \\ \sum_{k\neq j}^{n}\alpha_{jk}\phi_{k,1}\phi_{j,1}\phi_{k,n} & & & & & & \sum_{k\neq j}^{n}\alpha_{jk}\phi_{k,n}\phi_{j,n}\phi_{k,n} \end{pmatrix}_{n\times n^2} \begin{pmatrix} \Delta m_{11} \\ \Delta m_{21} \\ \cdots \\ \Delta m_{n1} \\ \cdots \\ \Delta m_{rs} \\ \cdots \\ \Delta m_{nn} \end{pmatrix}_{n^2\times 1}$$

$$- \frac{1}{2} \begin{pmatrix} \phi_{j,1}\phi_{j,1}\phi_{j,1} & \phi_{j,2}\phi_{j,1}\phi_{j,1} & \cdots & \phi_{j,n}\phi_{j,1}\phi_{j,1} & \cdots & \phi_{j,n}\phi_{j,n}\phi_{j,1} \\ \phi_{j,1}\phi_{j,1}\phi_{j,2} & \phi_{j,2}\phi_{j,1}\phi_{j,2} & & \phi_{j,n}\phi_{j,1}\phi_{j,2} & \cdots & \phi_{j,n}\phi_{j,n}\phi_{j,2} \\ \cdots & & & \cdots & \cdots & \\ \cdots & & & \cdots & \cdots & \\ \phi_{j,1}\phi_{j,1}\phi_{j,n} & & & \cdots & \phi_{j,n}\phi_{j,n}\phi_{j,n} \end{pmatrix}_{n\times n^2} \begin{pmatrix} \Delta m_{11} \\ \Delta m_{21} \\ \cdots \\ \Delta m_{n1} \\ \cdots \\ \Delta m_{rs} \\ \cdots \\ \Delta m_{nn} \end{pmatrix}_{n^2\times 1}.$$

(3.69)

Further defining the relevant terms in equation (3.69) as

$$\Phi_j = \Delta\phi_j = \begin{pmatrix} \Delta\phi_{j,1} \\ \Delta\phi_{j,2} \\ \cdots \\ \Delta\phi_{j,n} \end{pmatrix}_{n \times 1}, \tag{3.70}$$

$$P_j = \begin{pmatrix} \sum_{\substack{k=1 \\ k\neq j}}^{n}\alpha_{jk}\phi_{k,1}\phi_{j,1}\phi_{k,1} & \sum_{\substack{k=1 \\ k\neq j}}^{n}\alpha_{jk}\phi_{k,2}\phi_{j,1}\phi_{k,1} & \cdots & \sum_{\substack{k=1 \\ k\neq j}}^{n}\alpha_{jk}\phi_{k,n}\phi_{j,1}\phi_{k,1} & \cdots & \sum_{\substack{k=1 \\ k\neq j}}^{n}\alpha_{jk}\phi_{k,r}\phi_{j,s}\phi_{k,1} & \cdots & \sum_{\substack{k=1 \\ k\neq j}}^{n}\alpha_{jk}\phi_{k,n}\phi_{j,n}\phi_{k,1} \\ \sum_{\substack{k=1 \\ k\neq j}}^{n}\alpha_{jk}\phi_{k,1}\phi_{j,1}\phi_{k,2} & & & & & & & \\ \cdots & & & & & & & \\ \cdots & & & & \sum_{\substack{k=1 \\ k\neq j}}^{n}\alpha_{jk}\phi_{k,r}\phi_{j,s}\phi_{k,t} & & & \\ \cdots & & & & \cdots & & & \\ \sum_{\substack{k=1 \\ k\neq j}}^{n}\alpha_{jk}\phi_{k,1}\phi_{j,1}\phi_{k,n} & & & \cdots & & & \sum_{\substack{k=1 \\ k\neq j}}^{n}\alpha_{jk}\phi_{k,n}\phi_{j,n}\phi_{k,n} \end{pmatrix}_{n \times n^2} \tag{3.71}$$

$$Q_j = \begin{pmatrix} \sum_{\substack{k=1 \\ k\neq j}}^{n}\beta_{jk}\phi_{k,1}\phi_{j,1}\phi_{k,1} & \sum_{\substack{k=1 \\ k\neq j}}^{n}\beta_{jk}\phi_{k,2}\phi_{j,1}\phi_{k,1} & \cdots & \sum_{\substack{k=1 \\ k\neq j}}^{n}\beta_{jk}\phi_{k,n}\phi_{j,1}\phi_{k,1} & \cdots & \sum_{\substack{k=1 \\ k\neq j}}^{n}\beta_{jk}\phi_{k,r}\phi_{j,s}\phi_{k,1} & \cdots & \sum_{\substack{k=1 \\ k\neq j}}^{n}\beta_{jk}\phi_{k,n}\phi_{j,n}\phi_{k,1} \\ \sum_{\substack{k=1 \\ k\neq j}}^{n}\beta_{jk}\phi_{k,1}\phi_{j,1}\phi_{k,2} & & & & & & & \\ \cdots & & & & & & & \\ \cdots & & & & \sum_{\substack{k=1 \\ k\neq j}}^{n}\beta_{jk}\phi_{k,r}\phi_{j,s}\phi_{k,t} & & & \\ \cdots & & & & \cdots & & & \\ \sum_{\substack{k=1 \\ k\neq j}}^{n}\beta_{jk}\phi_{k,1}\phi_{j,1}\phi_{k,n} & & & \cdots & & & \sum_{\substack{k=1 \\ k\neq j}}^{n}\beta_{jk}\phi_{k,n}\phi_{j,n}\phi_{k,n} \end{pmatrix}_{n \times n^2}, \tag{3.72}$$

and

$$W_j = \frac{1}{2}\begin{pmatrix} \phi_{j,1}\phi_{j,1}\phi_{j,1} & \phi_{j,2}\phi_{j,1}\phi_{j,1} & \cdots & \phi_{j,n}\phi_{j,1}\phi_{j,1} & \cdots & \phi_{j,n}\phi_{j,n}\phi_{j,1} \\ \phi_{j,1}\phi_{j,1}\phi_{j,2} & \phi_{j,2}\phi_{j,1}\phi_{j,2} & & \phi_{j,n}\phi_{j,1}\phi_{j,2} & \cdots & \phi_{j,n}\phi_{j,n}\phi_{j,2} \\ \cdots & & & & \cdots & \cdots \\ \cdots & & & & \cdots & \cdots \\ \cdots & & & & \cdots & \cdots \\ \phi_{j,1}\phi_{j,1}\phi_{j,n} & & & \cdots & & \phi_{j,n}\phi_{j,n}\phi_{j,n} \end{pmatrix}_{n \times n^2}, \tag{3.73}$$

and substituting them into equation (3.69), the Φ_j, defined in (3.70), can finally be expressed in a compact matrix form as

$$\Phi_j = Q_j k - P_j m - W_j m. \tag{3.74}$$

The covariance matrix of Φ_j can therefore be derived as follows:

$$\begin{aligned} C_{\Phi_j} &= E(\Phi_j\Phi_j^T) - E[(Q_jk - P_jm - W_jm)(Q_jk - P_jm - W_jm)^T] \\ &= E[(Q_jk - P_jm - W_jm)(k^TQ_j^T - m^TP_j^T - m^TW_j^T)] \\ &= Q_jC_kQ_j^T + P_jC_mP_j^T + P_jC_mW_j^T + W_jC_mP_j^T + W_jC_mW_j^T. \end{aligned} \tag{3.75}$$

Again, C_{Φ_j} can be evaluated once C_k and C_m are known, and the matrices involved can be greatly reduced if there are only a small number of random elements in k and m (consequently C_k and C_m).

3.2.4 Derivation of the Covariance Matrix of the Defined Parameter $r_{jk,r}$

Considering the definition in equation (3.15), and assuming that changes being made to the defined parameter $r_{jk,r}$ are w.r.t. small changes in the spatial parameters, then[6]

$$r_{jk,r} + \Delta r_{jk,r} = (\phi_{r,j} + \Delta\phi_{r,j})F_k(\phi_{r,k} + \Delta\phi_{r,k}). \tag{3.76}$$

Expanding the above equation and ignoring the higher (second) order terms yields

$$\Delta r_{jk,r} \approx \phi_{r,j}F_k\Delta\phi_{r,k} + \Delta\phi_{r,j}F_k\phi_{r,k}. \tag{3.77}$$

Obviously equation (3.77) is a generic case of one element in the vector of the defined parameter Δr_{jk}. The whole vector can be deduced as

$$
\Delta r_{jk} =
\begin{pmatrix}
\phi_{1,j} & 0 & \cdots & 0 \\
0 & \phi_{2,j} & \cdots & \cdots \\
\cdots & \cdots & \cdots & \cdots \\
\cdots & \cdots & \cdots & \phi_{n,j}
\end{pmatrix}_{n\times n}
F_k
\begin{pmatrix}
\Delta\phi_{1,k} \\
\Delta\phi_{2,k} \\
\cdots \\
\Delta\phi_{n,k}
\end{pmatrix}_{n\times 1}
$$
$$
+
\begin{pmatrix}
\phi_{1,k} & 0 & \cdots & 0 \\
0 & \phi_{2,k} & \cdots & \cdots \\
\cdots & \cdots & \cdots & \cdots \\
\cdots & \cdots & \cdots & \phi_{n,k}
\end{pmatrix}_{n\times n}
F_k
\begin{pmatrix}
\Delta\phi_{1,j} \\
\Delta\phi_{2,j} \\
\cdots \\
\Delta\phi_{n,j}
\end{pmatrix}_{n\times 1}
. \tag{3.78}
$$

Defining the relevant terms in the above equation respectively as

$$
R_{jk} = \Delta r_{jk} =
\begin{pmatrix}
\Delta r_{jk,1} \\
\Delta r_{jk,2} \\
\cdots \\
\Delta r_{jk,n}
\end{pmatrix}_{n\times 1}
\tag{3.79}
$$

$$
S_j =
\begin{pmatrix}
\phi_{1,j} & 0 & \cdots & 0 \\
0 & \phi_{2,j} & \cdots & \cdots \\
\cdots & \cdots & \cdots & \cdots \\
\cdots & \cdots & \cdots & \phi_{n,j}
\end{pmatrix}_{n\times n}
\tag{3.80}
$$

---

6. It needs to be emphasized that, when multiple forces are applied, $r_{jk,r}$ will be complex. A newly defined parameter is needed, and is developed and presented in Appendix IV.

$$S_k = \begin{pmatrix} \phi_{1,k} & 0 & \cdots & 0 \\ 0 & \phi_{2,k} & \cdots & \cdots \\ \cdots & \cdots & \cdots & \cdots \\ \cdots & \cdots & \cdots & \phi_{n,k} \end{pmatrix}_{n \times n} \tag{3.81}$$

$$\tilde{\Phi}_j = \Delta \tilde{\phi}_j = \begin{pmatrix} \Delta\phi_{1,j} \\ \Delta\phi_{2,j} \\ \cdots \\ \Delta\phi_{n,j} \end{pmatrix}_{n \times 1}, \tag{3.82}$$

and

$$\tilde{\Phi}_k = \Delta \tilde{\phi}_k = \begin{pmatrix} \Delta\phi_{1,k} \\ \Delta\phi_{2,k} \\ \cdots \\ \Delta\phi_{n,k} \end{pmatrix}_{n \times 1}, \tag{3.83}$$

and substituting them into equation (3.78) yields

$$R_{jk} = F_k(S_j \Delta \tilde{\phi}_k + S_k \Delta \tilde{\phi}_j) = F_k(S_j \tilde{\Phi}_k + S_k \tilde{\Phi}_j). \tag{3.84}$$

It is important to appreciate the difference between $\Delta\phi_j(\Phi_j)$ and $\Delta \tilde{\phi}_j(\tilde{\Phi}_j)$. Their relationship is described in Figure 3.3.

As illustrated in Figure 3.3, $\Delta \tilde{\phi}_j$, defined in equation (3.82), is a vector consisting of all the elements in the jth row of a corresponding alternative of the mass-normalized mode shape matrix.[7]

$$\Delta\tilde{\phi}_j^T \} \begin{bmatrix} \Delta\phi_{1,j} & \cdots & \Delta\phi_{j,j} & \cdots & \Delta\phi_{n,j} \end{bmatrix} \begin{pmatrix} \Delta\psi_{1,1} & \cdots & \overbrace{\Delta\psi_{j,1}}^{\Delta\phi_j} & \cdots & \Delta\psi_{n,1} \\ \cdots & \cdots & \cdots & \cdots & \cdots \\ \Delta\phi_{1,j} & \cdots & \Delta\phi_{j,j} & \cdots & \Delta\phi_{n,j} \\ \cdots & \cdots & \cdots & \cdots & \cdots \\ \Delta\phi_{1,n} & \cdots & \underbrace{\Delta\phi_{j,n}} & \cdots & \Delta\phi_{n,n} \end{pmatrix}_{n \times n}$$

FIGURE 3.3   The definitions of $\Delta\phi_j$ and $\Delta\tilde{\phi}_j$.

7. The subscript used in Figure 3.3 is consistent with that for mode shape vectors, not for the whole mode shape matrix, which has no comma and the row number comes first.

Therefore, according to equation (3.69), $\Delta \tilde{\phi}_j$ can be assembled by selecting the corresponding elements in equation (3.69), yielding

$$
\Delta \tilde{\phi}_j = \begin{pmatrix} \Delta \phi_{1,j} \\ \Delta \phi_{2,j} \\ \cdots \\ \Delta \phi_{n,j} \end{pmatrix}_{N \times 1}
$$

$$
= \tilde{P}_j = \begin{pmatrix}
\sum_{k \neq 1}^{n} \beta_{1k} \phi_{k,1} \phi_{1,1} \phi_{k,j} & \sum_{k \neq 1}^{n} \beta_{1k} \phi_{k,2} \phi_{1,1} \phi_{k,j} & \cdots & \sum_{k \neq 1}^{n} \beta_{1k} \phi_{k,n} \phi_{1,1} \phi_{k,j} & \cdots & \sum_{k \neq 1}^{n} \beta_{1k} \phi_{k,r} \phi_{1,s} \phi_{k,j} & \cdots & \sum_{k \neq 1}^{n} \beta_{1k} \phi_{k,n} \phi_{1,n} \phi_{k,j} \\
\sum_{k \neq 2}^{n} \beta_{2k} \phi_{k,1} \phi_{2,1} \phi_{k,j} & \sum_{k \neq 2}^{n} \beta_{2k} \phi_{k,2} \phi_{2,1} \phi_{k,j} & & & & & & \sum_{k \neq 2}^{n} \beta_{2k} \phi_{k,n} \phi_{2,n} \phi_{k,j} \\
\cdots & \cdots & & & & & & \cdots \\
& & & & \sum_{k \neq j}^{n} \beta_{jk} \phi_{k,r} \phi_{j,s} \phi_{k,t} & & & \\
\cdots & & & & & & & \\
\sum_{k \neq n}^{n} \beta_{nk} \phi_{k,1} \phi_{n,1} \phi_{k,j} & \sum_{k \neq n}^{n} \beta_{nk} \phi_{k,2} \phi_{n,1} \phi_{k,j} & & & & & & \sum_{k \neq n}^{n} \beta_{nk} \phi_{k,n} \phi_{n,n} \phi_{k,j}
\end{pmatrix}_{n \times n^2} \begin{pmatrix} \Delta k_{11} \\ \Delta k_{21} \\ \cdots \\ \Delta k_{n1} \\ \cdots \\ \Delta k_{rs} \\ \cdots \\ \Delta k_{nn} \end{pmatrix}_{n^2 \times 1}
$$

$$
- \begin{pmatrix}
\sum_{k \neq 1}^{n} \alpha_{1k} \phi_{k,1} \phi_{1,1} \phi_{k,j} & \sum_{k \neq 1}^{n} \alpha_{1k} \phi_{k,2} \phi_{1,1} \phi_{k,j} & \cdots & \sum_{k \neq 1}^{n} \alpha_{1k} \phi_{k,n} \phi_{1,1} \phi_{k,j} & \cdots & \sum_{k \neq 1}^{n} \alpha_{1k} \phi_{k,r} \phi_{1,s} \phi_{k,j} & \cdots & \sum_{k \neq 1}^{n} \alpha_{1k} \phi_{k,n} \phi_{1,n} \phi_{k,j} \\
\sum_{k \neq 2}^{n} \alpha_{2k} \phi_{k,1} \phi_{2,1} \phi_{k,j} & \sum_{k \neq 2}^{n} \alpha_{2k} \phi_{k,2} \phi_{2,1} \phi_{k,j} & & & & & & \sum_{k \neq 2}^{n} \alpha_{2k} \phi_{k,n} \phi_{2,n} \phi_{k,j} \\
\cdots & \cdots & & & & & & \\
& & & & \sum_{k \neq j}^{n} \alpha_{jk} \phi_{k,r} \phi_{j,s} \phi_{k,t} & & & \\
\cdots & & & & & & & \\
\sum_{k \neq n}^{n} \alpha_{nk} \phi_{k,1} \phi_{n,1} \phi_{k,j} & \sum_{k \neq n}^{n} \alpha_{nk} \phi_{k,2} \phi_{n,1} \phi_{k,j} & & & & & & \sum_{k \neq n}^{n} \alpha_{nk} \phi_{k,n} \phi_{n,n} \phi_{k,j}
\end{pmatrix}_{n \times n^2} \begin{pmatrix} \Delta m_{11} \\ \Delta m_{21} \\ \cdots \\ \Delta m_{n1} \\ \cdots \\ \Delta m_{rs} \\ \cdots \\ \Delta m_{nn} \end{pmatrix}_{n^2 \times 1}
$$

$$
- \frac{1}{2} \begin{pmatrix}
\phi_{1,1} \phi_{1,1} \phi_{1,j} & \phi_{1,2} \phi_{1,1} \phi_{1,j} & \cdots & \phi_{1,n} \phi_{1,1} \phi_{1,j} & \cdots & \phi_{1,n} \phi_{1,n} \phi_{1,j} \\
\phi_{2,1} \phi_{2,1} \phi_{2,j} & \phi_{2,2} \phi_{2,1} \phi_{2,j} & & \phi_{2,n} \phi_{2,1} \phi_{2,j} & \cdots & \phi_{2,n} \phi_{2,n} \phi_{2,j} \\
\cdots & & \cdots & \cdots & & \\
\cdots & & \cdots & \cdots & & \\
\cdots & & \cdots & \cdots & & \\
\phi_{n,1} \phi_{n,1} \phi_{n,j} & & & \cdots & \phi_{n,n} \phi_{n,n} \phi_{n,j}
\end{pmatrix}_{n \times n^2} \begin{pmatrix} \Delta m_{11} \\ \Delta m_{21} \\ \cdots \\ \Delta m_{n1} \\ \cdots \\ \Delta m_{rs} \\ \cdots \\ \Delta m_{nn} \end{pmatrix}_{n^2 \times 1}
$$

(3.85)

Defining the relevant terms in the above equation respectively as

$$
\tilde{P}_j = \begin{pmatrix}
\sum_{k \neq 1}^{n} \alpha_{1k} \phi_{k,1} \phi_{1,1} \phi_{k,j} & \sum_{k \neq 1}^{n} \alpha_{1k} \phi_{k,2} \phi_{1,1} \phi_{k,j} & \cdots & \sum_{k \neq 1}^{n} \alpha_{1k} \phi_{k,n} \phi_{1,1} \phi_{k,j} & \cdots & \sum_{k \neq 1}^{n} \alpha_{1k} \phi_{k,r} \phi_{1,s} \phi_{k,j} & \cdots & \sum_{k \neq 1}^{n} \alpha_{1k} \phi_{k,n} \phi_{1,n} \phi_{k,j} \\
\sum_{k \neq 2}^{n} \alpha_{2k} \phi_{k,1} \phi_{2,1} \phi_{k,j} & \sum_{k \neq 2}^{n} \alpha_{2k} \phi_{k,2} \phi_{2,1} \phi_{k,j} & & & & & & \sum_{k \neq 2}^{n} \alpha_{2k} \phi_{k,n} \phi_{2,n} \phi_{k,j} \\
\cdots & \cdots & & & & & & \cdots \\
& & & & \sum_{k \neq j}^{n} \alpha_{jk} \phi_{k,r} \phi_{j,s} \phi_{k,t} & & & \\
\cdots & & & & & & & \\
\sum_{k \neq n}^{n} \alpha_{nk} \phi_{k,1} \phi_{n,1} \phi_{k,j} & \sum_{k \neq n}^{n} \alpha_{nk} \phi_{k,2} \phi_{n,1} \phi_{k,j} & & & & & & \sum_{k \neq n}^{n} \alpha_{nk} \phi_{k,n} \phi_{n,n} \phi_{k,j}
\end{pmatrix}_{n \times n^2}
$$

(3.86)

$$
\tilde{Q}_j = \begin{pmatrix}
\sum_{k \neq 1}^{n} \beta_{1k} \phi_{k,1} \phi_{1,1} \phi_{k,j} & \sum_{k \neq 1}^{n} \beta_{1k} \phi_{k,2} \phi_{1,1} \phi_{k,j} & \cdots & \sum_{k \neq 1}^{n} \beta_{1k} \phi_{k,n} \phi_{1,1} \phi_{k,j} & \cdots & \sum_{k \neq 1}^{n} \beta_{1k} \phi_{k,r} \phi_{1,s} \phi_{k,j} & \cdots & \sum_{k \neq 1}^{n} \beta_{1k} \phi_{k,n} \phi_{1,n} \phi_{k,j} \\
\sum_{k \neq 2}^{n} \beta_{2k} \phi_{k,1} \phi_{2,1} \phi_{k,j} & \sum_{k \neq 2}^{n} \beta_{2k} \phi_{k,2} \phi_{2,1} \phi_{k,j} & & & & & & \sum_{k \neq 2}^{n} \beta_{2k} \phi_{k,n} \phi_{2,n} \phi_{k,j} \\
\cdots & \cdots & & & & & & \cdots \\
& & & & \sum_{k \neq j}^{n} \beta_{jk} \phi_{k,r} \phi_{j,s} \phi_{k,t} & & & \\
\cdots & & & & & & & \\
\sum_{k \neq n}^{n} \beta_{nk} \phi_{k,1} \phi_{n,1} \phi_{k,j} & \sum_{k \neq n}^{n} \beta_{nk} \phi_{k,2} \phi_{n,1} \phi_{k,j} & & & & & & \sum_{k \neq n}^{n} \beta_{nk} \phi_{k,n} \phi_{n,n} \phi_{k,j}
\end{pmatrix}_{n \times n^2} ,
$$

(3.87)

and

$$\tilde{W}_j = \frac{1}{2}\begin{pmatrix} \phi_{1,1}\phi_{1,1}\phi_{1,j} & \phi_{1,2}\phi_{1,1}\phi_{1,j} & \cdots & \phi_{1,n}\phi_{1,1}\phi_{1,j} & \cdots & \phi_{1,n}\phi_{1,n}\phi_{1,j} \\ \phi_{2,1}\phi_{2,1}\phi_{2,j} & \phi_{2,2}\phi_{2,1}\phi_{2,j} & & \phi_{2,n}\phi_{2,1}\phi_{2,j} & \cdots & \phi_{2,n}\phi_{2,n}\phi_{2,j} \\ \cdots & & & & & \\ \cdots & & & & \cdots & \cdots \\ \cdots & & & & \cdots & \cdots \\ \phi_{n,1}\phi_{n,1}\phi_{n,j} & & & & \cdots & \phi_{n,n}\phi_{n,n}\phi_{n,j} \end{pmatrix}_{n \times n^2},$$

$$(3.88)$$

and substituting them into equation (3.85) results in

$$\tilde{\Phi}_j = \tilde{Q}_j k - \tilde{P}_j m - \tilde{W}_j m \tag{3.89}$$

and

$$\tilde{\Phi}_k = \tilde{Q}_k k - \tilde{P}_k m - \tilde{W}_k m. \tag{3.90}$$

Substituting equations (3.89) and (3.90) into equation (3.84) yields

$$R_{jk} = F_k[S_j(\tilde{Q}_k k - \tilde{P}_k m - \tilde{W}_k m) + S_k(\tilde{Q}_j k - \tilde{P}_j m - \tilde{W}_j m)]. \tag{3.91}$$

Finally, the covariance matrix for R_{jk} can be obtained as follows:

$$\begin{aligned} C_{R_{jk}} &= F_k^2 E[(S_j\tilde{\Phi}_k + S_k\tilde{\Phi}_j)(S_j\tilde{\Phi}_k + S_k\tilde{\Phi}_j)^T] \\ &= F_k^2 E[(S_j\tilde{\Phi}_k + S_k\tilde{\Phi}_j)(\tilde{\Phi}_k^T S_j^T + \tilde{\Phi}_j^T S_k^T)] \\ &= F_k^2 E[S_j\tilde{\Phi}_k\tilde{\Phi}_k^T S_j^T + S_k\tilde{\Phi}_j\tilde{\Phi}_k^T S_j^T + S_j\tilde{\Phi}_k\tilde{\Phi}_j^T S_k^T + S_k\tilde{\Phi}_j\tilde{\Phi}_j^T S_k^T] \\ &= F_k^2 [S_j E(\tilde{\Phi}_k\tilde{\Phi}_k^T)S_j^T + S_k E(\tilde{\Phi}_j\tilde{\Phi}_k^T)S_j^T + S_j E(\tilde{\Phi}_k\tilde{\Phi}_j^T)S_k^T + S_k E(\tilde{\Phi}_j\tilde{\Phi}_j^T)S_k^T] \\ &= F_k^2 [S_j C_{\tilde{\Phi}_k}S_j^T + S_k E(\tilde{\Phi}_j\tilde{\Phi}_k^T)S_j^T + S_j E(\tilde{\Phi}_k\tilde{\Phi}_j^T)S_k^T + S_k C_{\tilde{\Phi}_j}S_k^T]. \end{aligned}$$

$$(3.92)$$

Because

$$\begin{aligned} E(\tilde{\Phi}_j\tilde{\Phi}_k^T) &= E[(\tilde{Q}_j k - \tilde{P}_j m - \tilde{W}_j m)(\tilde{Q}_k k - \tilde{P}_k m - \tilde{W}_k m)^T] \\ &= E[(\tilde{Q}_j k - \tilde{P}_j m - \tilde{W}_j m)(k^T\tilde{Q}_k^T - m^T\tilde{P}_k^T - m^T\tilde{W}_k^T)] \\ &= \tilde{Q}_j C_k\tilde{Q}_k^T + \tilde{P}_j C_m\tilde{P}_k^T + \tilde{P}_j C_m\tilde{W}_k^T + \tilde{W}_j C_m\tilde{P}_k^T + \tilde{W}_j C_m\tilde{W}_k^T \end{aligned} \tag{3.93}$$

and

$$E(\tilde{\Phi}_k\tilde{\Phi}_j^T) = \tilde{Q}_k C_k\tilde{Q}_j^T + \tilde{P}_k C_m\tilde{P}_j^T + \tilde{P}_k C_m\tilde{W}_j^T + \tilde{W}_k C_m\tilde{P}_j^T + \tilde{W}_k C_m\tilde{W}_j^T$$

$$(3.94)$$

$$C_{\Phi_j} = E(\tilde{\Phi}_j \tilde{\Phi}_j^T) = \tilde{Q}_j C_k \tilde{Q}_j^T + \tilde{P}_j C_m \tilde{P}_j^T + \tilde{P}_j C_m \tilde{W}_j^T + \tilde{W}_j C_m \tilde{P}_j^T + \tilde{W}_j C_m \tilde{W}_j^T,$$

$$(3.95)$$

and

$$C_{\Phi_k} = E(\tilde{\Phi}_k \tilde{\Phi}_k^T) = \tilde{Q}_k C_k \tilde{Q}_k^T + \tilde{P}_k C_m \tilde{P}_k^T + \tilde{P}_k C_m \tilde{W}_k^T + \tilde{W}_k C_m \tilde{P}_k^T + \tilde{W}_k C_m \tilde{W}_k^T.$$

$$(3.96)$$

Equation (3.92) can be rewritten in terms of C_k and C_m as

$$C_{R_{jk}} = F_k^2 \begin{bmatrix} S_j(\tilde{Q}_k C_k \tilde{Q}_k^T + \tilde{P}_k C_m \tilde{P}_k^T + \tilde{P}_k C_m \tilde{W}_k^T + \tilde{W}_k C_m \tilde{P}_k^T + \tilde{W}_k C_m \tilde{W}_k^T)S_j^T \\ + S_k(\tilde{Q}_j C_k \tilde{Q}_k^T + \tilde{P}_j C_m \tilde{P}_k^T + \tilde{P}_j C_m \tilde{W}_k^T + \tilde{W}_j C_m \tilde{P}_k^T + \tilde{W}_j C_m \tilde{W}_k^T)S_j^T \\ + S_j(\tilde{Q}_k C_k \tilde{Q}_j^T + \tilde{P}_k C_m \tilde{P}_j^T + \tilde{P}_k C_m \tilde{W}_j^T + \tilde{W}_k C_m \tilde{P}_j^T + \tilde{W}_k C_m \tilde{W}_j^T)S_k^T \\ + S_k(\tilde{Q}_j C_k \tilde{Q}_j^T + \tilde{P}_j C_m \tilde{P}_j^T + \tilde{P}_j C_m \tilde{W}_j^T + \tilde{W}_j C_m \tilde{P}_j^T + \tilde{W}_j C_m \tilde{W}_j^T)S_k^T \end{bmatrix}.$$

$$(3.97)$$

Considering that the S matrices are symmetric, i.e. $S_j^T = S_j$, $S_k^T = S_k$, $C_{R_{jk}}$ can be finally written as

$$C_{R_{jk}} = F_k^2 \begin{bmatrix} S_j(\tilde{Q}_k C_k \tilde{Q}_k^T + \tilde{P}_k C_m \tilde{P}_k^T + \tilde{P}_k C_m \tilde{W}_k^T + \tilde{W}_k C_m \tilde{P}_k^T + \tilde{W}_k C_m \tilde{W}_k^T)S_j \\ + S_k(\tilde{Q}_j C_k \tilde{Q}_k^T + \tilde{P}_j C_m \tilde{P}_k^T + \tilde{P}_j C_m \tilde{W}_k^T + \tilde{W}_j C_m \tilde{P}_k^T + \tilde{W}_j C_m \tilde{W}_k^T)S_j \\ + S_j(\tilde{Q}_k C_k \tilde{Q}_j^T + \tilde{P}_k C_m \tilde{P}_j^T + \tilde{P}_k C_m \tilde{W}_j^T + \tilde{W}_k C_m \tilde{P}_j^T + \tilde{W}_k C_m \tilde{W}_j^T)S_k \\ + S_k(\tilde{Q}_j C_k \tilde{Q}_j^T + \tilde{P}_j C_m \tilde{P}_j^T + \tilde{P}_j C_m \tilde{W}_j^T + \tilde{W}_j C_m \tilde{P}_j^T + \tilde{W}_j C_m \tilde{W}_j^T)S_k \end{bmatrix}.$$

$$(3.98)$$

3.2.5 Derivation of the Covariance Matrix of the Combined Parameter T

It is expected that the defined parameters d_r and $r_{jk,r}$ are correlated with each other. To apply the FORM method, it is necessary to de-correlate the random variables before the normalization is processed. Therefore, it is appropriate to develop a parameter combining the two variables that contains the covariance information.

The combined parameter is defined as T and is written as

$$T_{jk} = \begin{pmatrix} D \\ R_{jk} \end{pmatrix}_{2nx1} = (\Delta d_1 \quad \Delta d_2 \quad \ldots \quad \Delta d_n \quad \Delta r_{jk,1} \quad \Delta r_{jk,2} \quad \ldots \quad \Delta r_{jk,n})_{2n \times 1}^T.$$

$$(3.99)$$

Substituting equation (3.37) into equation (3.51) results in

$$D = HAk - HBm. \tag{3.100}$$

Substituting equations (3.100) and (3.91) into equation (3.99) yields

$$T_{jk} = \left(\begin{matrix} HA \\ F_k(S_j \tilde{Q}_k + S_k \tilde{Q}_j) \end{matrix} \right)_{2n \times n^2} k - \left(\begin{matrix} HB \\ F_k(S_j \tilde{P}_k + S_j \tilde{W}_k + S_k \tilde{P}_j + S_k \tilde{W}_j) \end{matrix} \right)_{2n \times n^2} m. \tag{3.101}$$

Defining the two terms in equation (3.101) respectively as

$$U_{jk} = \left(\begin{matrix} HA \\ F_k(S_j \tilde{Q}_k + S_k \tilde{Q}_j) \end{matrix} \right)_{2n \times n^2} \tag{3.102}$$

and

$$V_{jk} = \left(\begin{matrix} HB \\ F_k(S_j \tilde{P}_k + S_j \tilde{W}_k + S_k \tilde{P}_j + S_k \tilde{W}_j) \end{matrix} \right)_{2n \times n^2}, \tag{3.103}$$

the alternative matrix of the combined parameter can be described as

$$T_{jk} = U_{jk}k - V_{jk}m. \tag{3.104}$$

The covariance matrix of T can therefore be derived as

$$\begin{aligned} C_{T_{jk}} &= E(T_{jk}T_{jk}^T) = E[(U_{jk}k - V_{jk}m)(U_{jk}k - V_{jk}m)^T] \\ &= E[(U_{jk}k - V_{jk}m)(k^T U_{jk}^T - m^T V_{jk}^T)] \\ &= U_{jk}C_k U_{jk}^T + V_{jk}C_m V_{jk}^T. \end{aligned} \tag{3.105}$$

3.2.6 Derivation of the Mean Values of the Defined Parameters d_r and $r_{jk,r}$

Considering the expanded equation (3.42) of the d_r definition equation, the expected value of d_r can be written as

$$F(d_r) = F(\omega_r^4 - 2\omega_r^2\omega_r^2 + \omega^4 + \eta_r^2\omega_r^4) = F(\omega_r^4) - 2\omega^2 F(\omega_r^2) + \omega^4 + \eta_r^2 F(\omega_r^4) \tag{3.106}$$

Applying the statistics rules results in

$$E(d_r) = (1 + \eta_r^2)[V(\omega_r^2) + E(\omega_r^2)^2] - 2\omega^2 E(\omega_r^2) + \omega^4, \tag{3.107}$$

where V denotes the "variance of".

The term $E(\omega_r^2)$ in the above equation, i.e. the expected value (mean) of ω_r^2, can be determined by eigen solution of the mean values of K and M. $V(\omega_r^2)$ can be obtained from the covariance matrix C_D in equation (3.54), or $C_{T_{jk}}$ in equation (3.105), the rth elements on the diagonal of these matrices. Other terms are known values.

Applying the statistics rules to the definition equation (3.15) of $r_{jk,r}$ results in

$$E(r_{jk,r}) = F_k[E(\phi_{r,j})E(\phi_{r,k}) + Cov(\phi_{r,j}, \phi_{r,k})], \qquad (3.108)$$

where $Cov(\phi_{r,j}, \phi_{r,k})$ is the covariance of the two elements $\phi_{r,j}$ and $\phi_{r,k}$. Similarly, $E(\phi_{r,j})$ and $E(\phi_{r,k})$ can be determined by taking the eigen solution on the mean value of K and M. $Cov(\phi_{r,j}, \phi_{r,k})$ can be obtained from the covariance matrix C_{Φ_j} in equation (3.75).

Thus, so far, once the covariance matrix of the spatial parameters is known, the first two moments of the defined variables d_r and $r_{jk,r}$ can then be evaluated through the algebraic algorithms developed in this section.

All the important definitions and results of the perturbation analysis are summarized and presented in Appendix V.

3.3 APPLICATION PROCEDURE OF THE NEW APPROACH

The procedure of the proposed reliability approach integrating the FE method and a probabilistic method, for example the FORM method, can now be outlined below:

1. Assembling the relevant mass and stiffness matrices M and K using the mean values of any random variable involved.
2. Finding the eigenvectors (the mass-normalized mode shapes $\{\Phi\} = [\{\phi_1\}, \{\phi_2\}, \ldots, \{\phi_n\}]$) and eigenvalues (the natural frequency squares $\omega_1^2, \omega_2^2, \ldots, \omega_n^2$) by solving eig$(K,M)$.
3. Assembling the covariance matrices of C_k and C_m, according to the known random variables.
4. Assembling all the relevant intermediate matrices, for example A, B, S, P, Q, W, U, V, according to the definitions given in the last section.
5. Assembling the defined covariance matrices, such as $C_{T_{jk}}$, to gather all the relevant statistical information on the defined random variables.
6. Applying the FORM method, de-correlating d_r and $r_{jk,r}$[8] according to the information obtained in $C_{T_{jk}}$.
7. Transforming every random variable d_r or $r_{jk,r}$ into their normalized forms, according to the new moment values calculated.
8. Construct the safety margin in the normalized space.
9. Applying the FORM iteration algorithm equation (2.19) to obtain the safety index β and then the probability of failure by $\Phi(-\beta)$.

The accuracy and efficiency of the new approach have been investigated by two case studies, which will be presented in the following two chapters.

---

8. The subscript mode number r should have been identified in advance.

3.4 DISCUSSION

The proposed approach is a novel reliability technique that is based on an approximate modal model and relevant perturbation analysis. As mentioned in Chapter 1, the modal superposition model was also adopted by Moens, Vandepitte and colleagues [91−97] to analyze structural dynamics with uncertainties, but with different techniques. The difference between the Moens−Vandepitte approach and the proposed approach has been reviewed and briefly discussed here.

In Refs [91,94,97,98], for undamped structures, the FRF between nodes j and k was expressed as

$$\text{FRF}_{jk} = \sum_{r=1}^{n} \frac{\phi_{r,j}\phi_{r,k}}{\{\phi_r\}^T[K]\{\phi_r\} - \omega^2\{\phi_r\}^T[M]\{\phi_r\}}. \tag{3.109}$$

FRF_{jk} was further written as

$$\text{FRF}_{jk} = \sum_{r=1}^{n} \frac{1}{D(\omega)}, \tag{3.110}$$

where $D(\omega) = \hat{k}_r - \omega^2\hat{m}_r$ is the modal response denominator, and $\hat{k}_r$ and $\hat{m}_r$ are defined modal parameters, which have the form:

$$\hat{k}_r = \frac{\{\phi_r\}^T[K]\{\phi_r\}}{\phi_{r,j}\phi_{r,k}}, \hat{m}_r = \frac{\{\phi_r\}^T[M]\{\phi_r\}}{\phi_{r,j}\phi_{r,k}}.$$

Proportional damping was assumed; therefore, the system damping matrix $[C]$ can be expressed as

$$[C] = \alpha_K[K] + \alpha_M[M], \tag{3.111}$$

where α_K and α_M are proportional damping coefficients. The deterministic damped FRF finally becomes

$$\text{FRF}_{jk} = \sum_{r=1}^{n} \frac{1}{(\hat{k}_r - \omega^2\hat{m}_r) + i\omega(\alpha_K\hat{k}_r - \alpha_M\hat{m}_r)}. \tag{3.112}$$

The procedure for the fuzzy finite element FRF analysis is:

1. Assemble the fuzzy mass and stiffness matrices respectively from the fuzzy inputs and determine the modal parameter range vector ($\hat{k}_r$ and $\hat{m}_r$).
2. Calculate the modal envelope FRF by substituting the ranges of the modal parameters in the denominator function $D(\omega)$ and determine the real and imaginary parts of equation (3.112).
3. Obtain the total interval FRF by the summation of the contribution of each mode.

The differences between the proposed approach and that of Moens and Vandepitte are summarized in Table 3.1.

TABLE 3.1 Differences Between the Two Approaches

Differences/ Approach	Moens and Vandepitte's Approach	Proposed Approach
Modal model	Deterministic modal superposition model $$\text{FRF}_{jk} = \sum_{r=1}^{n} \frac{1}{D(\omega)}$$	Approximate modal model $$\left|\overline{X}_j\right|^2 \approx \sum_{r=1}^{n} \frac{\left|\phi_{r,j} F_k \phi_{r,k}\right|^2}{(\omega_r^2 - \omega^2)^2 + (\eta_r \omega_r^2)^2}$$
Damping mechanism	Proportional damping	Hysteretic damping
Defined modal parameter	$$\hat{k}_r = \frac{\{\phi_r\}^T [K]\{\phi_r\}}{\phi_{r,j}\phi_{r,k}}, \hat{m}_r = \frac{\{\phi_r\}^T [M]\{\phi_r\}}{\phi_{r,j}\phi_{r,k}}$$	$$r_{jk,r} = \phi_{r,j} F_k \phi_{r,k}$$ $$d_r = (\omega_r^2 - \omega^2)^2 + (\eta_r \omega_r^2)^2$$
Analysis technique	Fuzzy interval method	Probabilistic methods, e.g. FORM, MCS

3.5 SUMMARY

The technical fundamentals of the proposed approach have been presented in this chapter. The novelty of the work and the conditions are summarized as below:

1. An approximated modal model has been developed, based on the primary interests in resonance cases and the condition of low modal overlap factor (low-frequency range). Equation (3.13) is the approximation result determined by neglecting all but the most significant terms at resonance in the modal model represented by equation (3.10). The approximation accuracy of the new model that mimics the resonance responses will be analyzed in the next chapter.

2. A set of new parameters has been introduced to replace the original random variables in order to overcome the nonlinearity problem of the failure surface in applying reliability methods such as the FORM method. The denominator and numerator of equation (3.13) were respectively defined as two new random variables. The linearity property of the new failure surface (based on this parameter transformation) was presented by the discussions of equations (3.17) and (3.18). Further analysis will be conducted through case studies in the following chapters.

3. The statistical properties of the new variables have been developed from those of the original variables and the modal properties (natural frequencies and mode shapes). The detailed derivations of the covariance matrices and

the mean values of the defined parameters have been presented in Section 3.2. It was assumed that the covariance matrices of the mass and stiffness matrices are known in advance. The required natural frequencies and mode shapes can be obtained by taking just once the eigen solution of the mean values of the mass and stiffness properties. This would improve the efficiency of the new approach compared to the full Monte Carlo simulations with a great number of FE calls and eigen solutions. This will be tested in the next chapters.

4. Small changes in eigenvalues are a condition for the perturbation approach to work, as linear approximations were applied in the derivations of the statistical moments of the defined parameters.

Chapter 4

Application to a 2D System

In Chapter 3, the theoretical fundamentals of the proposed combined approach were introduced. Developed as a reliability analysis solution for dynamic systems, the approach integrates a novel modal model-based perturbation technique and one of the probabilistic methods. In-depth case studies have been conducted to apply and validate the approach on two dynamic systems with *added masses*. The first one is a two-dimensional frame structure. However, theoretical and technical problems were found during this study and solutions were sought and applied, which are presented in this chapter.

4.1 FINITE ELEMENT MODEL OF A 2D DYNAMIC SYSTEM

The example system is chosen from Ref. [116, p. 258]. As illustrated in Figure 4.1, it is a right-angle frame structure, which is uniform, pinned at the point A and roller-supported at C, where a load is applied horizontally. The geometry and constraining information is also given in the figure. The property values, which are treated as *deterministic* parameters, are listed in Table 4.1.

The structure is modeled by Femlab™ finite element software. As illustrated in Figure 4.2, the FE model has five nodes and four elements.[1] Since this is a two-dimensional model, each node has three DOFs (in x, y, and in-plane rotation), and thus the system has in total 15 DOFs. However, as three of them are constrained, the final total number of DOFs is reduced to 12. All DOF information is presented in Table 4.2. A load is applied on node 5 in the x direction. As shown in Figure 4.1, two added masses are placed on nodes 2 and 4 respectively, representing any equipment or supporting objects that might be implemented in real cases. The reliability concern is focused on the horizontal (x-direction) displacement of node 2 for analyzing the dynamic effect on the equipment.

As introduced in earlier chapters, the spatial model, i.e. the mass matrix $[M]$ and the stiffness matrix $[K]$, can be assembled by the FE process. They are all 12 by 12 matrices. The modal model, i.e. the *natural frequencies* and *mode shapes*, can be determined by solving the eigen problem.[2] They are a 12-element vector and a 12 by 12 matrix respectively.

---

1. It should be noted that this FE model of the structure is greatly simplified. The main purpose is to demonstrate the application of the combined approach rather than to build a realistic model, which requires many more nodes and elements.
2. The corresponding Matlab command is eig(K,M).

Reliability Analysis of Dynamic Systems.
© 2013 Shanghai Jiao Tong University Press. Published by Elsevier Inc. All rights reserved.

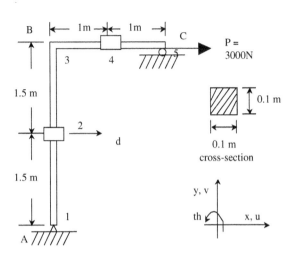

FIGURE 4.1 A two-dimensional right-angle frame structure (with FE node numbers).

TABLE 4.1 Property Values of the Structure

Property	E	ν	G	ρ	I	P	η
Values	200×10^9 Pa	0.29	77.5×10^9 Pa	7860 kg/m^3	$0.1^4/12$	3000 N	0.05

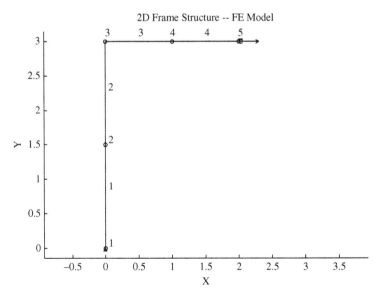

FIGURE 4.2 FE model of the 2D frame structure (with the node and element numbers).

TABLE 4.2 DOF Indices of the FE Model (15 DOFs in Total)

	u (x)		v (y)		θ (Rotation)	
Node 1	1 (fixed)	×	6 (fixed)	×	11	8
Node 2	2 (random mass 1 and displacement)	1	7	5	12	9
Node 3	3	2	8	6	13	10
Node 4	4	3	9 (random mass 2)	7	14	11
Node 5	5 (force applied)	4	10 (fixed)	×	15	12

TABLE 4.3 Natural Frequencies (in Hz) of the Structure

Mode No.	Natural Frequency (Original)	Natural Frequency (with Mean Added Masses)
1	3.31	2.49
2	35.25	18.11
3	71.78	29.01
4	139.72	131.79
5	260.90	238.26
6	365.60	323.48
7	418.47	396.59
8	564.97	520.04
9	937.88	887.69
10	1249.59	1248.50
11	1375.77	1371.60
12	2579.32	2574.70

All the natural frequencies of the system with and without the added masses, in hertz, are listed in Table 4.3 and the first four mode shapes are illustrated in Figure 4.3.[3]

---

3. The higher mode shapes shown in Figure 4.3 were performed by Femlab with a default parameter called *element refinement* set to 3 that adds more plotting points to produce smoother shapes.

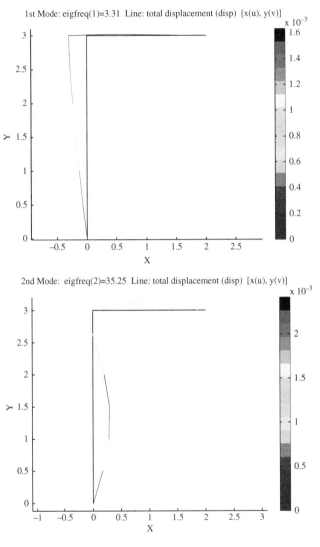

FIGURE 4.3 First four original mode shapes of the 2D dynamic system.

Since all the physical parameters are assumed to be deterministic, the assembled stiffness matrix $[K]$ is therefore deterministic. So is the original mass matrix $[M]$. It is the *two added masses* that are assumed to be random. Practically, this is often the case for the equipment or objects represented.

3rd Mode: eigfreq(3)=71.78 Line: total displacement (disp) [x(u), y(v)]

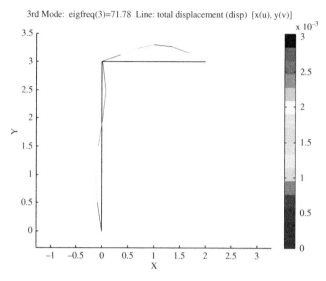

4th Mode: eigfreq(4)=139.71 Line: total displacement (disp) [x(u), y(v)]

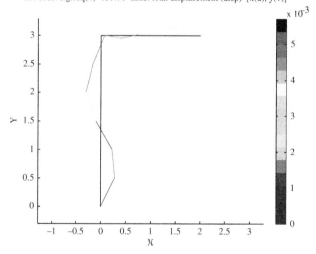

FIGURE 4.3   *(Continued)*

The two added random masses, m_1 and m_2, are assumed to be Gaussian, both with a mean value of 500 kg.

In the mass matrix $[M]$, there are three mass elements for each node, corresponding to each DOF. Thus in total six elements would be influenced by the two random added masses. To simplify the issue, it is assumed that

2nd Mode (added mass): eigfreq(2)=18.15 Line: total displacement (disp) [x(u), y(v)]

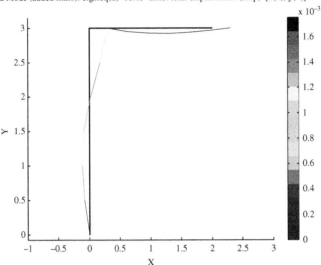

FIGURE 4.4 The second mode shape of the dynamic system with added masses.

only the elements in the transverse direction, i.e. one on node 2 horizontally (x-direction) and another on node 4 vertically (y-direction), are random.[4]

According to Table 4.2, the DOF index for the horizontal displacement of node 2 is 1, and that for the load on node 5 is 4. The excitation frequency is assumed to be around the *second natural frequency*, i.e. 18.11 Hz, to mimic the resonance response. Therefore, the subscripts j, k and r in the defined parameters $r_{jk,r}$ and d_r in equations (3.14) and (3.15) can now be determined, i.e. $j = 1$, $k = 4$, and $r = 2$. The second mode shape with the mean added masses is demonstrated in Figure 4.4. The numerical details of all the 12 mode shapes are presented in Table 4.4.

4.2 APPLYING THE COMBINED APPROACH: PRELIMINARY ANALYSIS

The new approach is an approximate solution. It is therefore important to perform a *response analysis* on the new approach in comparison with the exact FE solution to validate the accuracy of the approximation. In addition, transforming the safety margin from closed curves to monotonic surfaces is a

4. For example, due to a fluid loading. Also note that by adopting the random masses, the size of the covariance matrix of the mass matrix, C_m, will be greatly reduced (with only the non-zero items left). The calculations described in Chapter 3 will consequently be greatly simplified because the size of the other matrices involved will also be correspondingly reduced. See also the discussion under equation (3.75).

TABLE 4.4 Numerical Details of All 12 Mode Shapes with Added Masses

R/C	1	2	3	4	5	6	7	8	9	10	11	12
1	0.0295	0.0228	0.0160	−0.0048	−0.0079	−0.0059	−0.0002	0.0001	0.0015	−0.0005	0.0012	−0.0005
2	0.0465	−0.0417	−0.0323	0.0173	−0.0118	−0.0110	−0.0003	0.0038	−0.0188	−0.0917	0.0749	−0.0896
3	0.0465	−0.0418	−0.0324	0.0180	−0.0135	−0.0146	−0.0005	0.0104	0.0179	0.0132	−0.0014	0.1334
4	0.0465	−0.0418	−0.0324	0.0183	−0.0142	−0.0158	−0.0005	0.0130	0.0364	0.0953	−0.0749	−0.1491
5	0.0000	0.0001	−0.0003	0.0003	0.0053	−0.0153	−0.0543	0.0231	−0.0197	−0.0543	−0.0825	0.0065
6	0.0000	0.0001	−0.0006	0.0006	0.0097	−0.0254	−0.0813	0.0273	−0.0022	0.0464	0.0937	−0.0179
7	−0.0021	0.0254	−0.0330	0.0028	−0.0011	−0.0049	0.0048	0.0052	−0.0040	−0.0003	0.0001	−0.0006
8	−0.0212	−0.0325	−0.0323	−0.3103	−0.3164	−0.3937	−0.0309	−0.5058	−0.1586	0.0290	−0.0818	0.0260
9	−0.0166	0.0181	0.0292	0.2502	−0.0390	−0.2188	−0.0242	−0.5078	−0.1965	0.0383	−0.1082	0.0358
10	−0.0056	0.0455	−0.0138	−0.1153	0.3321	0.1624	−0.0071	−0.5238	−0.2643	0.3316	−0.4108	0.3268
11	0.0007	−0.0022	−0.0099	0.0438	−0.2937	0.3154	−0.1315	0.0411	−0.8481	0.4701	−0.0110	0.1436
12	0.0028	−0.0371	0.0550	−0.0318	0.2737	−0.4712	0.3841	0.4301	−1.0820	0.4507	−0.0097	0.1119

fundamental feature of the new approach, which also needs to be validated before the FORM method is applied.

4.2.1 Response Analysis

The responses at node 2 were analyzed in terms of three different approaches, namely the *FE solution*, *modal solution* and the defined *approximation solution*, which are discussed below:

1. *FE solution*, by inverting the dynamic stiffness matrix (referring to equations (AIII.15 and AIII.23))

$$\{\overline{X}\} = [-\omega^2 M + (1+i\eta)K]^{-1}\{F\}.\tag{4.1}$$

Taking the square function[5] on the *j*th individual node displacement of (4.1) yields

$$|\overline{X}_j|^2 = |[-\omega^2 M + (1+i\eta)K]^{-1}\{F\}_j|^2.\tag{4.2}$$

2. *Modal solution*, using equation (3.9),[6]

$$|\overline{X}_j|^2 = \sum_{r=1}^{N}\sum_{p=1}^{N} \frac{(r_{jk,r})(r_{jk,p})^*}{\left[(\omega_r^2 - \omega^2) + i\eta_r\omega_r^2\right]\left[(\omega_p^2 - \omega^2) - i\eta_p\omega_p^2\right]}.\tag{4.3}$$

3. *Approximation solution*, using equation (3.16),

$$|\overline{X}_j|^2 = \sum_{r=1}^{N} \frac{r_{jk,r}^2}{d_r},\tag{4.4}$$

where

$$r_{jk,r} = \phi_{r,j}F_k\phi_{r,k}\tag{4.5}$$

$$d_r = (\omega_r^2 - \omega^2)^2 + (\eta_r\omega_r^2)^2.\tag{4.6}$$

The responses against varying excitation frequencies, with two different η values, i.e. 0.05 and 0.01, were generated by a Matlab program according to the above three solutions. The common logarithms of the responses of the first three modes (0 to 40 Hz) are presented in Figures 4.5 and 4.6.

---

5. This is to remain consistent with the approximation solution.
6. The results of the modal solution are the same as those of the FE solution. The modal solution is added to the response analysis to relate the FE solution to the modal summation equations (from which the approximation solution is derived). Hereafter, in the following reliability analysis, the displacement determined by the FE solution will be used as the exact reference for comparison with other methods.

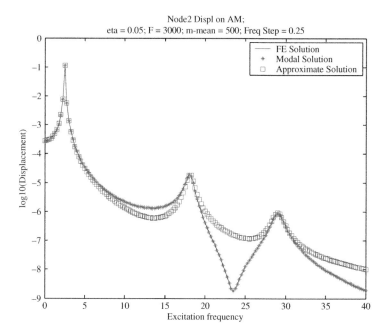

FIGURE 4.5   Response analysis of node 2 vs. excitation frequencies, $\eta = 0.05$.

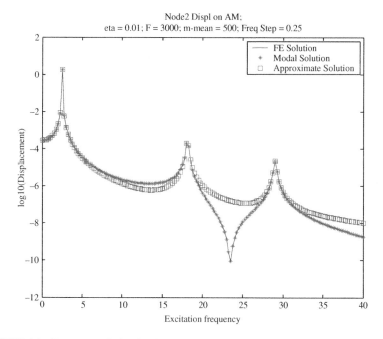

FIGURE 4.6   Response analysis of node 2 vs. excitation frequencies, $\eta = 0.01$.

According to the results, the resonance occurs when the excitation frequency coincides with the natural frequencies.

The responses generated by the approximation solution match very well to that of the "exact" solution (FE solution or modal solution) at the resonance peak, particularly when the *damping loss factor* is low (as shown in the second figure, the response obtained by the approximation solution with $\eta = 0.01$ is closer to the FE solution than in the first figure). This has proved that *the proposed perturbation approach is accurate in approximating the resonance displacement*. In the later analysis, to validate that the proposed approach works in the "worst" case, η was taken to be 0.05.

4.2.2 Safety Margin Contour

The definition of the safety margin on node 2 involves the difference between a displacement threshold, known as the *maximum allowed displacement* denoted as D_{Max}, and the "real" displacement obtained by either the proposed combined approach or the reference one, i.e. the FE solution.

The random variables in the FE solution are the two added masses, while the random variables in the approximation solution are the defined parameters r and d (equation (3.16)). Since it is assumed that the excitation frequency is around the second natural frequency (mode 2), r_2 and d_2 are therefore chosen as the random variables for the reliability analysis, because mode 2 is the dominant one, as explained by equation (3.18). The remaining r and d terms in equation (3.16), corresponding to other modes, are assumed to be deterministic (using their expected values). Given that j, k, and r have been determined earlier, and assuming that the loss factors have the same η for the whole system, the definitions of r_2 and d_2 can now be written as

$$r_2 = \phi_{2,1} F_4 \phi_{2,4} \tag{4.7}$$

$$d_2 = (\omega_2^2 - \omega^2)^2 + (\eta \omega_2^2)^2. \tag{4.8}$$

The definitions of the two safety margins are listed in Table 4.5. For consistency with the response analysis, the square of the response of node 2 and D_{Max} were found, and then their common logarithms were taken.

As illustrated in Figure 4.4, the resonance response (a maximum) under the excitation frequency at mode 2 ($f = 18.11$ Hz, $\eta = 0.05$, and mean added mass $= 500$ kg) is about 0.0044 m. Therefore, two values of D_{Max} are suggested, one within the failure region,[7] for example 0.003 m, another very close to the failure surface,[8] i.e. 0.0044 m. The safety margin contours, with

---

7. The probability of failure can be obtained by $1 - \Phi(-\beta) = \Phi(\beta)$.

8. The reason for choosing these two interesting values is to test the approach under some special circumstances. According to the practice of FORM discussed in Chapter 2, if it works in these two cases, it should work in other cases where the probability of failure is significantly lower.

TABLE 4.5 Safety Margin Definitions of Different Analysis Approaches

Approach	Definition of Safety Margin
FE solution (inverting the dynamic stiffness matrix)	$\log\lvert D_{\text{Max}}\rvert^2 - \log\lvert\{[-\omega^2 M + (1+i\eta)K]^{-1}F\}_1\rvert^2$
Combined approach (perturbation algorithm to get statistical information of R and D + probabilistic method)	$\log\lvert D_{\text{Max}}\rvert^2 - \log\left\lvert\dfrac{r_2^2}{d_2} + \displaystyle\sum_{i=1,i\neq2}^{12}\dfrac{E(r_i)^2}{E(d_i)}\right\rvert$

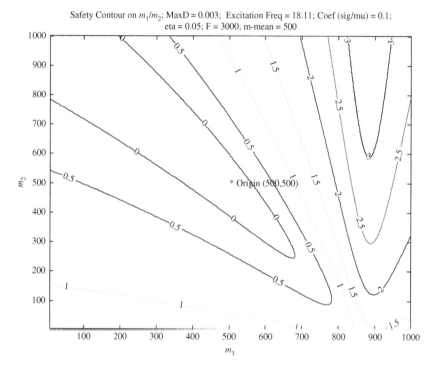

Safety Contour on m_1/m_2; MaxD = 0.003; Excitation Freq = 18.11; Coef (sig/mu) = 0.1;
eta = 0.05; F = 3000; m-mean = 500

FIGURE 4.7   Safety margin contour on two added masses m_1 and m_2, $D_{\text{Max}} = 0.003$.

these two D_{Max} values and the two random mass variables, are presented in Figures 4.7 and 4.8 respectively.

Clearly, both contours show that the safety margins are in the form of closed and non-monotonic curves. As discussed in Chapter 3, this problem can be overcome if the safety margin is constructed on the

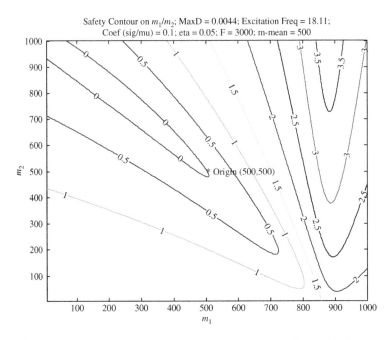

FIGURE 4.8   Safety margin contour on two added masses m_1 and m_2, $D_{Max} = 0.0044$.

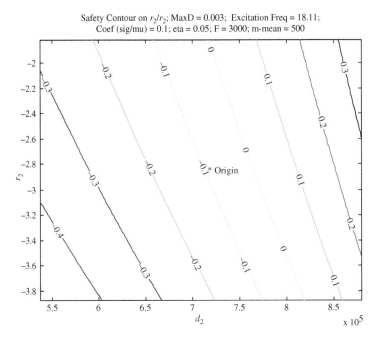

FIGURE 4.9   Safety margin contour on two defined parameters r_2 and d_2, $D_{Max} = 0.003$.

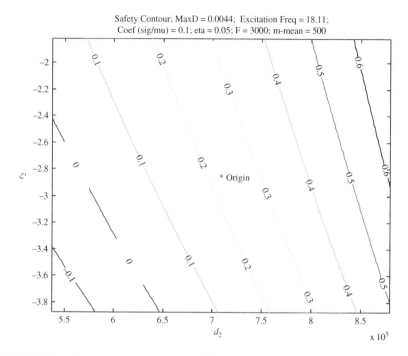

FIGURE 4.10   Safety margin contour on two defined parameters r_2 and d_2, $D_{Max} = 0.0044$.

defined parameters R and D. This has been proved, according to Figures 4.9 and 4.10.

This transformation is essential for application of the FORM method. The required statistical information of the defined parameters can be obtained using the perturbation algorithms presented in Chapter 3.

4.3 PERTURBATION APPROACH + FORM METHOD

As discussed before, the purpose of the perturbation approach is to apply the probabilistic reliability methods, such as FORM. The outcome of the preliminary analysis complies with what is expected. This section deals with the reliability application of this combined approach.

4.3.1 Evaluating the Probability of Failure and In-Depth Analysis

A Matlab computer program has been developed to apply the combined approach, i.e. perturbation approach + FORM method, as well as the reference Monte Carlo simulation method on FE solutions.

A description of the relevant pseudo computer code is outlined below.

Perturbation + FORM (Proposed Approach)

- Assemble the mass matrix $[M]$ with the mean added masses, m_1 and m_2.
- Call the eigen solver eig(K,M) only once to obtain all the relevant modal properties, such as natural frequencies and mode shapes.
- Assemble the covariance matrix of $[M]$, C_m, and all other relevant perturbation matrices that do not involve the excitation frequency.
- Loop through excitation frequencies (e.g. from 16.5 to 19.5 Hz), and for each excitation frequency:
  - Apply the relevant perturbation algorithm to determine the mean values of r_2 and according to equations (3.107) and (3.108).*
  - Apply the relevant perturbation algorithm to obtain C_T, which contains the variance information of r_2 and d_2.
  - Apply the FORM method, according to the procedure given in Section 3.3, obtain the safety index β and then the probability of failure, Prob_F (currentFreq) $= \Phi(-\beta)$.
- End of excitation frequency loop.

* r_2 is not dependent on any excitation frequency, thus it can be evaluated outside the frequency loop.

Two different displacement thresholds and different coefficients of variation (ratio of the standard deviation and the mean) of the random mass variables, denoted as *Coef*, are chosen to run the program, i.e.

I. $D_{\text{Max}} = 0.003$, *Coef* $= 0.1$.
II. $D_{\text{Max}} = 0.0044$, *Coef* $= 0.05$.

Other parameters are fixed, e.g. $\eta = 0.05$, $F = 3000$ (for the reason for these two settings, please refer to footnote 8).

According to the two settings, the probabilities of failure under different excitation frequencies were calculated and plotted. These are presented in Figures 4.11 and 4.12 respectively.

Monte Carlo Simulation on FE Solution ("Exact" Solution used as a Reference)

- Loop through excitation frequencies (e.g. from 16.5 to 19.5 Hz), and for each excitation frequency:
  - Set *failure_number* to 0.
  - Loop through the number of MC trials until TotalMC is reached.
    - Find a pair of realization of the random variables of m_1 and m_2 (using the known mean and standard deviation).
  - Assemble the mass matrix $[M]$ with the two added masses.

- Process the FE solution, i.e. solve equation (4.1), or use the modal solution, i.e. solve the eigen problem eig(K,M) and evaluate equation (4.3) to obtain the displacement value on node 2, i.e. $\overline{X}_1$.*
- Evaluate the defined safety margin given in Table 4.5; if it is less than 0, which represents "failed", increase the *failure_number* by 1.
- End of Monte Carlo simulation loop.
- Obtain the probability of failure under the current excitation frequency:

  Prob_F(currentFreq) = *failure_number*/TotalMC.

- End of excitation frequency loop.

* As proved before, both methods yields the same results, though the modal solution may be more efficient for a large system as it does not need to process matrix inverse calculations as the FE solution does.

Significant discrepancies were observed between the Perturbation + FORM approach and the Monte Carlo solution in predicting the probability of failure, which suggests an inaccurate approximation was made by the combined approach.

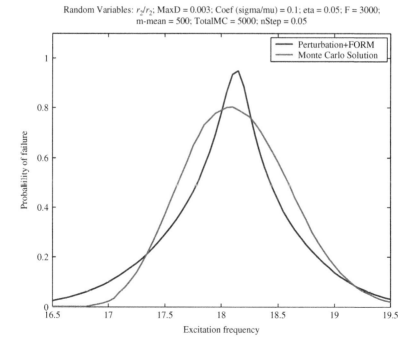

Random Variables: r_2/r_2; MaxD = 0.003; Coef (sigma/mu) = 0.1; eta = 0.05; F = 3000; m-mean = 500; TotalMC = 5000; nStep = 0.05

FIGURE 4.11   Reliability analysis of r_2 and d_2, $D_{Max} = 0.003$, $Coef = 0.1$.

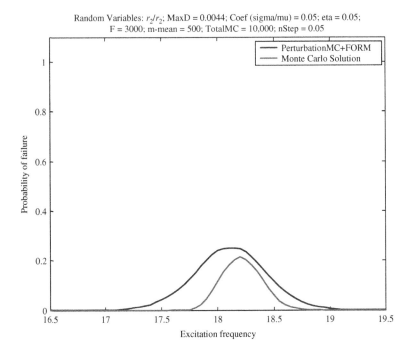

Random Variables: r_2/r_2; MaxD = 0.0044; Coef (sigma/mu) = 0.05; eta = 0.05; F = 3000; m-mean = 500; TotalMC = 10,000; nStep = 0.05

FIGURE 4.12 Reliability analysis of r_2 and d_2, $D_{\text{Max}} = 0.0044$, $Coef = 0.05$.

To locate the problem, approximation analyses were first carried out on r_2, d_2 and their elements to compare the *mean, standard deviation* and *variation* (the ratio of standard deviation and mean) predicted by the perturbation approach with those simulated by the Monte Carlo method. This was performed using the same program procedure described earlier. Instead of the probability of failure, the mean, standard deviation and variation were calculated, under varying excitation frequencies.

Approximation Analysis of r_2 and its Components

Two mode shape elements involved in r_2, i.e. $\phi_{2,1}$ and $\phi_{2,4}$, were analyzed first.

The mode shape elements including $\phi_{2,1}$ and $\phi_{2,4}$ are determined by the eigen solution, which does not depend on any excitation frequency. Neither does r_2. Means of $\phi_{2,1}$ and $\phi_{2,4}$ were predicted by the perturbation approach and by a 5000 times Monte Carlo simulation. As shown in Figure 4.13, two pairs of results are presented together under the same scale, each of them containing two comparable lines that are almost identical.

Further analysis of r_2, which is a function of the product of the two mode shape elements $\phi_{2,1}$ and $\phi_{2,4}$ (augmented by the force amplitude), reveals that there is little difference between the new approach and the Monte Carlo simulation in predicting r_2. According to the plots presented in Figures 4.14–4.16,

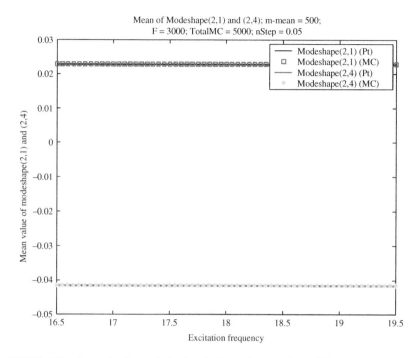

FIGURE 4.13   Approximation analysis of mode shape elements, $\phi_{2,1}$ and $\phi_{2,4}$.

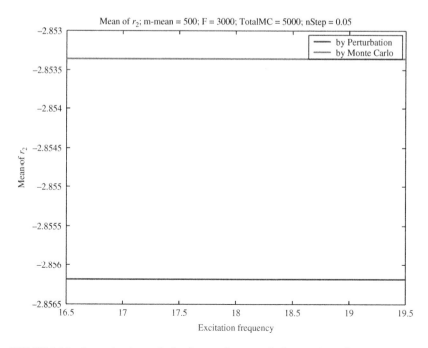

FIGURE 4.14   Approximation analysis of mean of r_2, perturbation vs. Monte Carlo.

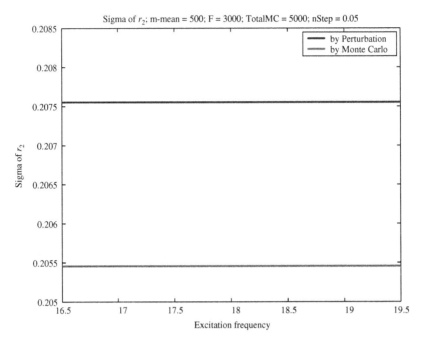

FIGURE 4.15 Approximation analysis of standard deviation of r_2, perturbation vs. Monte Carlo.

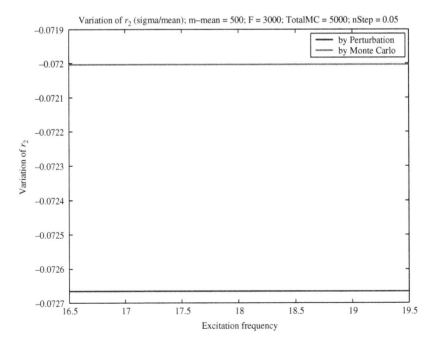

FIGURE 4.16 Approximation analysis of variation of r_2, perturbation vs. Monte Carlo.

the discrepancies between the mean, the standard deviation and the coefficient of variation[9] of r_2 obtained by the two approaches can be ignored.

Conclusion: The mean and variance of r_2 are accurately predicted by the new approach.

Approximation Analysis of d_2 and its Components

Similarly, the approximation analysis of d_2, which is defined in equation (4.6), starts from its only random component ω_2^2. Plots of the corresponding mean, standard deviation and variation of ω_2^2 produced by the perturbation approach and the MC method are presented in Figures 4.17, 4.18, and 4.19 respectively.

Being one of the system's natural features (modal characteristics), ω_2^2 does not depend on any excitation frequency or external forces, but the added masses. Figures 4.17–4.19 prove that the evaluation of ω_2^2 performed by the perturbation approach (by taking only the mean values of the added masses in the eigen solution) is close enough to that by a number of Monte Carlo simulations (the error is below 0.5%).

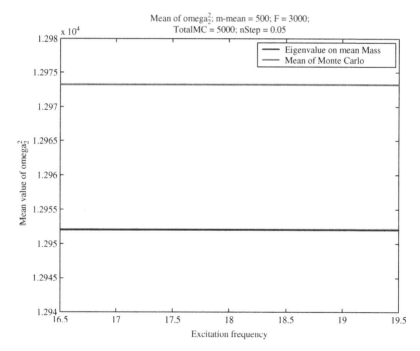

FIGURE 4.17   Approximation analysis of mean of ω_2^2, perturbation vs. Monte Carlo.

9. The coefficient of variation is defined as σ/μ and abbreviated in the figure captions as "variation".

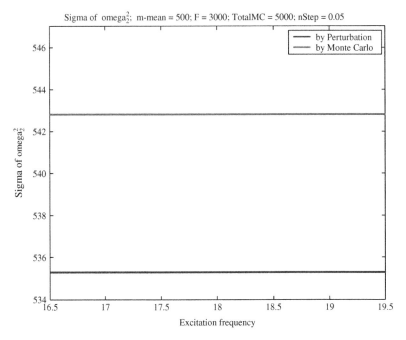

FIGURE 4.18   Approximation analysis of standard deviation of ω_2^2, perturbation vs. Monte Carlo.

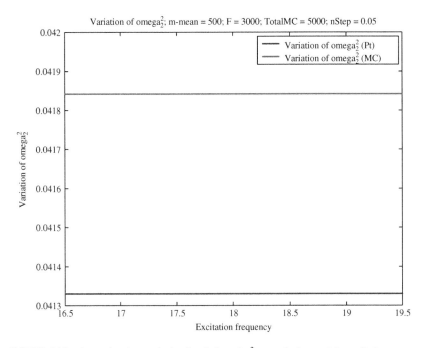

FIGURE 4.19   Approximation analysis of variation of ω_2^2, perturbation vs. Monte Carlo.

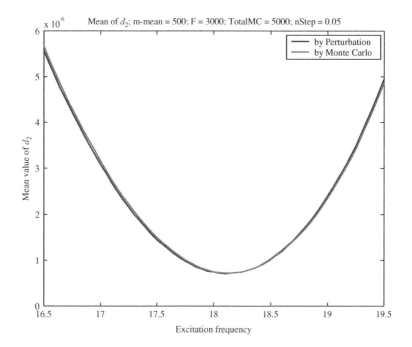

FIGURE 4.20   Approximation analysis of mean of d_2, perturbation vs. Monte Carlo.

However, further analysis of d_2 showed that there is an obvious difference between the two approaches (though the two mean values are matched well, by Figure 4.20), particularly in evaluating the standard deviation and variation of d_2, according to Figures 4.21 and 4.22. This might be due to the fact that d_2 was only approximated linearly, i.e. only the first-order terms remain in equation (3.47).

Conclusion: The new approach is not accurate in approximating the coefficient of variation of d_2, especially at the resonance where the approximation accuracy is particularly poor. The equations for the approximation of d_2, i.e. equations (3.42)–(3.47), need to be reviewed.

4.3.2 Solution 1: Second-Order Approximation of d_2

Approximation Analysis of d_2

It was intuitive to speculate that the problem found in the previous section might be an inaccurate approximation of d_2 in equation (3.47). The full terms in equation (3.46) were therefore reviewed and the second-order term was added to replace the linear approximation.

Adding the ignored second-order term $\Delta\omega_r^4\left((\Delta\omega_r^2)^2\right)$ to equation (3.47) results in

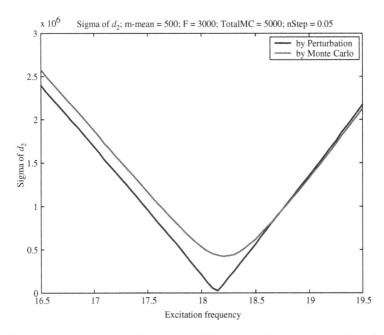

FIGURE 4.21 Approximation analysis of standard deviation of d_2, perturbation vs Monte Carlo.

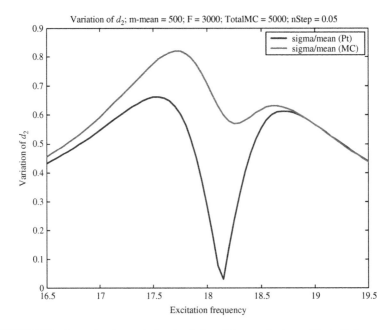

FIGURE 4.22 Approximation analysis of variation of d_2, perturbation vs Monte Carlo.

$$\Delta d_r = 2(\omega_r^2 - \omega^2 + \eta_r^2 \omega_r^2)\Delta\omega_r^2 + (1 + \eta_r^2)\Delta\omega_r^4. \tag{4.9}$$

Letting $\alpha = 2(\omega_r^2 - \omega^2 + \eta_r^2 \omega_r^2)$ and $X = \Delta\omega_r^2$, and substituting them into the above equation, yields

$$\Delta d_r = \alpha X + (1 + \eta_r^2)X^2. \tag{4.10}$$

Squaring on both sides generates

$$(\Delta d_r)^2 = \alpha^2 X^2 + 2\alpha(1 + \eta_r^2)X^3 + (1 + \eta_r^2)^2 X^4. \tag{4.11}$$

Because the parameter X, i.e. $\Delta\omega_r^2$, is assumed to be Gaussian, the following statistical rules (properties) apply:

$$\begin{aligned} E[X] &= 0 \\ E[X^2] &= \sigma_X^2 = V(X) \\ E[X^3] &= 0 \\ E[X^4] &= 3\sigma_X^4 = 3V^2(X), \end{aligned} \tag{4.12}$$

where $E[X]$ is the expected value of X, $V(X)$ is the variance of X, and σ_X is the standard deviation of X.

Therefore, the new *variance* of Δd_r, denoted as $V(\Delta d_r)$, can be derived as follows (expanding equation (4.11) and applying (4.12)):

$$\begin{aligned} V(\Delta d_r) = E[(\Delta d_r)^2] &= E[\alpha^2 X^2 + 2\alpha(1 + \eta_r^2)X^3 + (1 + \eta_r^2)^2 X^4] \\ &= \alpha^2 E[X^2] + 2\alpha(1 + \eta_r^2)E[X^3] + (1 + \eta_r^2)^2 E[X^4] \\ &= \alpha^2 V(X) + 3(1 + \eta_r^2)^2 V^2(X). \end{aligned} \tag{4.13}$$

Replacing X with $\Delta\omega_r^2$ in equation (4.13) finally gives

$$V(\Delta d_r) = \alpha^2 V(\Delta\omega_r^2) + 3(1 + \eta_r^2)^2 V^2(\Delta\omega_r^2). \tag{4.14}$$

$V(\Delta\omega_r^2)$ are the diagonal elements of the *covariance matrix* C_Ω. Therefore, the new variance of Δd_r can be evaluated once C_Ω is known, which can be determined by the perturbation algorithm.

Finally, the new standard deviation of d_2 can be written as

$$\sigma_{d_2} = \sqrt{V(\Delta d_2)} - \sqrt{\alpha^2 V(\Delta\omega_2^2) + 3(1 + \eta_2^2)^2 V(\Delta\omega_2^2)^2}. \tag{4.15}$$

The old and new variations of d_2, i.e. σ_d/μ_d, obtained by the perturbation approach are plotted along with that obtained by the Monte Carlo simulations in Figure 4.23.

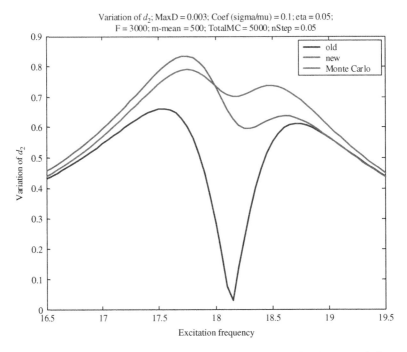

FIGURE 4.23 Approximation analysis of variation of d_2 (second order), perturbation vs. Monte Carlo.

Clearly, the new variation of d_2, with the second-order approximation, is much closer to the "exact" value.

Reliability Analysis by the New Approximation of d_2

Consequently, this enhancement was added to the perturbation + FORM approach. The second-order approximation of d_2, equation (4.15), was used to replace the earlier linear prediction. Two reliability cases were re-analyzed and the results are presented in Figures 4.24 and 4.25.

Though the accuracy has been improved in predicting the probability of failure by adding the second-order term to the approximation of d_2, discrepancies are still apparent compared to the "exact" solution in each case. Further analysis is needed.

As discussed in Chapter 2, there are two necessary conditions for the FORM method to be accurate, i.e. (1) all the random variables are Gaussian and (2) the linear approximation to the failure surface must be reasonable. The closer to these conditions, the better the approximation will be. Condition (2) has been checked earlier in this chapter (see Figures 4.9 and 4.10). It is now

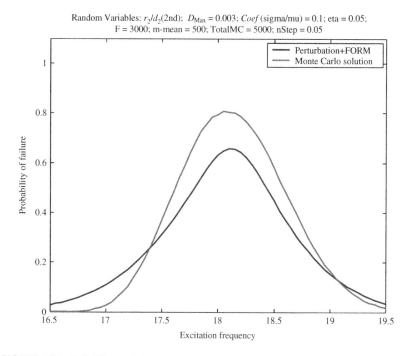

Random Variables: r_2/d_2(2nd); D_{Max} = 0.003; $Coef$ (sigma/mu) = 0.1; eta = 0.05; F = 3000; m-mean = 500; TotalMC = 5000; nStep = 0.05

FIGURE 4.24 Reliability analysis of r_2 and d_2 (second order), D_{Max} = 0.003, $Coef$ = 0.1.

reasonable to suspect that either or both of the defined parameters are *not Gaussian*.

Conclusion: While adding the second-order term definitely improved the approximation of d_2, there was still a significant inaccuracy problem when predicting the probability of failure by the combined approach. This requires further analysis to find whether or not the random variables involved, i.e. r_2 and d_2, are normally distributed.

Normal Distribution Analysis of r_2 and d_2

To cover some typical cases, the analysis was performed at three different excitation frequencies, i.e. the resonance frequency 18.1 Hz, and two non-resonance frequencies 16.5 and 19.5 Hz (either side of the resonance peak). For each case, two plots were generated. The first one was a *histogram plot*,[10] which bins all the Monte Carlo simulations of r_2 and d_2 into 20 equally spaced containers. It aims to show a direct distribution pattern of the tested random variable. The second was a *normal probability plot*,[11]

---

10. The Matlab command is "hist".
11. The Matlab command is "normplot".

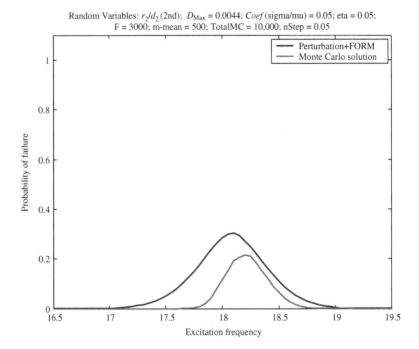

FIGURE 4.25 Reliability analysis of r_2 and d_2 (second order), $D_{Max} = 0.0044$, $Coef = 0.05$.

which graphically assesses whether the data in the random variable would obey a normal distribution. If the data were normal the plot would be linear. Other distribution types would introduce curvature to the plot.

Normal Distribution Analysis of r_2

According to the six plots of r_2, i.e. histogram plots and normal probability plots for 16.5, 18.1 and 19.5 Hz (presented in Figures 4.26−4.31), it can readily be concluded that the random variable r_2 is approximately Gaussian.

Normal Distribution Analysis of d_2

In contrast, d_2 is not Gaussian, according to the six plots of d_2 presented in Figures 4.32−4.37. Particularly when the system is excited under the resonance frequency 18.11 Hz, d_2 exhibits a strongly non-Gaussian distribution (Figures 4.34 and 4.35), which may be due to the higher order of the elements involved in the definition of d_2.

Conclusion: The defined parameter r_2 is Gaussian while d_2 is not. In order to apply the FORM method, a new parameter is needed to replace d_2.

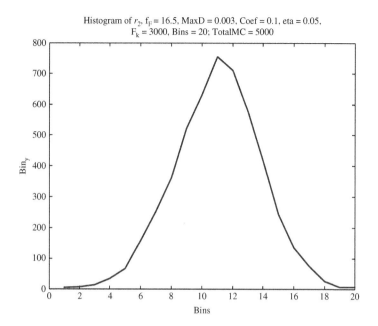

FIGURE 4.26   Histogram of $r_2, f_F = 16.5$, bins $= 20$.

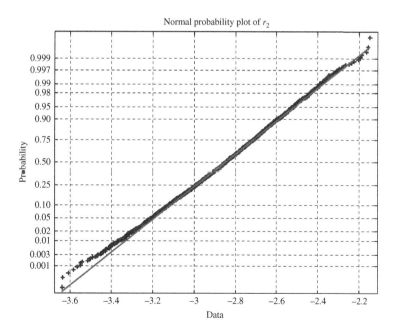

FIGURE 4.27   Normal probability of $r_2, f_F = 16.5$, bins $= 20$.

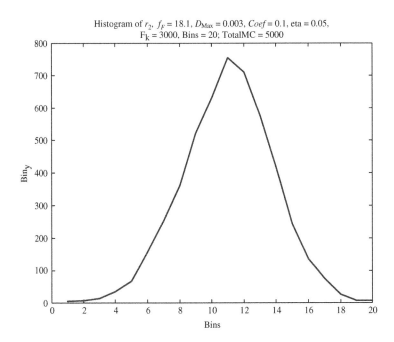

FIGURE 4.28   Histogram of r_2, $f_F = 18.1$, bins = 20.

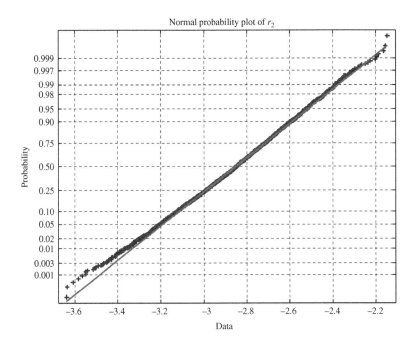

FIGURE 4.29   Normal probability of r_2, $f_F = 18.1$, bins = 20.

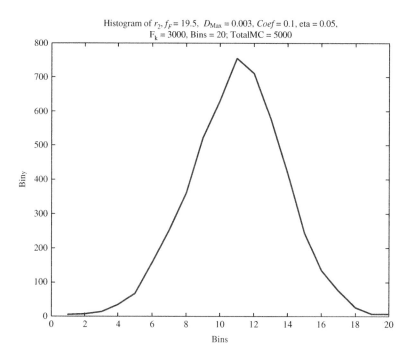

FIGURE 4.30   Histogram of r_2, $f_F = 19.5$, bins = 20.

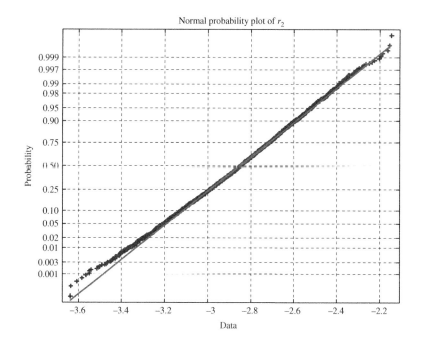

FIGURE 4.31   Normal probability of r_2, $f_F = 19.5$, bins = 20.

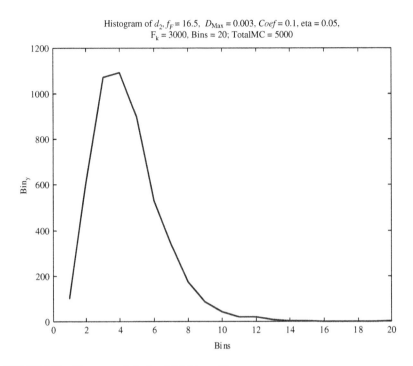

FIGURE 4.32   Histogram of d_2, $f_F = 16.5$, bins = 20.

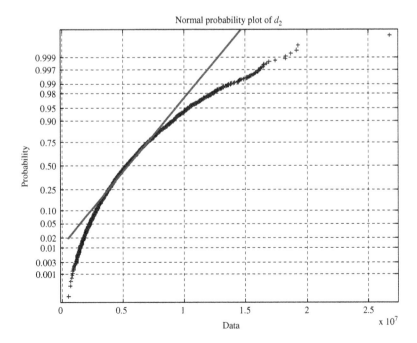

FIGURE 4.33   Normal probability of d_2, $f_F = 16.5$, bins = 20.

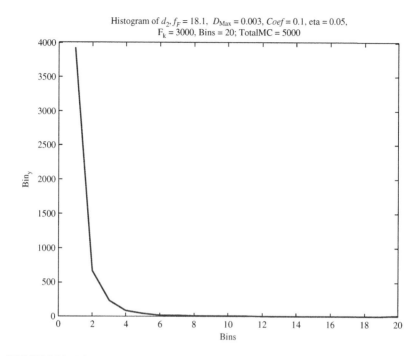

FIGURE 4.34   Histogram of d_2, $f_F = 18.1$, bins = 20.

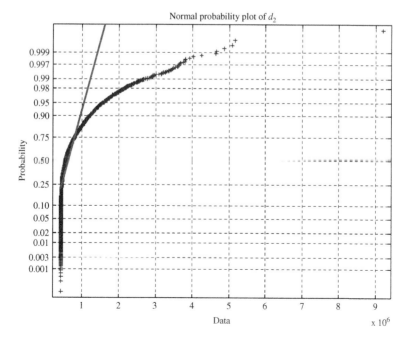

FIGURE 4.35   Normal probability of d_2, $f_F = 18.1$, bins = 20.

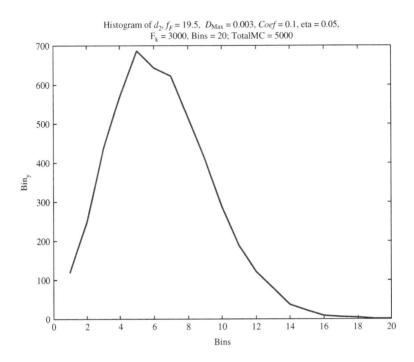

FIGURE 4.36 Histogram of d_2, $f_F = 19.5$, bins $= 20$.

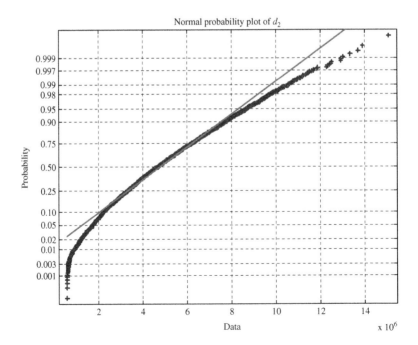

FIGURE 4.37 Normal probability of d_2, $f_F = 19.5$, bins $= 20$.

4.3.3 Solution 2: New Variable e_2 to Replace d_2

To solve the non-Gaussian problem, a new parameter, named e_2, is introduced to replace the problematic d_2. e_2 is suggested to have the following relationship with d_2:[12]

$$e_2 = \sqrt{d_2} \tag{4.16}$$

or

$$d_2 = e_2^2. \tag{4.17}$$

The required mean and standard deviation of e_2 can be derived as follows. Being random variables, d_2 and e_2 can be written as

$$d_2 = \mu_d + \hat{d} \tag{4.18}$$

$$e_2 = \mu_e + \hat{e}, \tag{4.19}$$

where μ_d and μ_e are the means of d_2 and e_2 respectively, $\hat{d}$ and $\hat{e}$ are the random errors.

Assuming that e_2 is Gaussian, it has the following properties:

$$\begin{aligned} E[\hat{e}] &= 0 \\ E[\hat{e}^2] &= \sigma_e^2 = V(e) \\ E[\hat{e}^3] &= 0 \\ E[\hat{e}^4] &= 3\sigma_e^4 = 3V^2(e), \end{aligned} \tag{4.20}$$

where σ_e and $V(e)$ are the standard deviation and the variance of e_2 respectively.

Substituting (4.19) into the definition equation (4.17) yields d_2 in terms of e_2:

$$d_2 = e_2^2 = (\mu_e + \hat{e})^2 = \mu_e^2 + 2\mu_e\hat{e} + \hat{e}^2 \tag{4.21}$$

and the expected value of d_2 (by applying (4.20)) is

$$E[d_2] = \mu_d = \mu_e^2 + \sigma_e^2. \tag{4.22}$$

Substituting equation (4.19) into the square of the expression for d_2 yields

$$d_2^2 = (\mu_e + \hat{e})^4 = \mu_e^4 + 4\mu_e^3\hat{e} + 6\mu_e^2\hat{e}^2 + 4\mu_e\hat{e}^3 + \hat{e}^4. \tag{4.23}$$

Taking the expected value of d_2^2 and applying (4.20) results in

$$E[d_2^2] = \mu_e^4 + 6\mu_e^2\sigma_e^2 + 3\sigma_e^4. \tag{4.24}$$

---

12. According to equation (3.14), at resonance, i.e. when $w_2 \simeq w$, $d_2 \simeq (\eta w_2^2)^2$. Thus the square root of d_2 at resonance would have the same statistical property as w_2^2 (the eigenvalue), which would be close to Gaussian.

Because

$$E[d_2^2] = \mu_d^2 + \sigma_d^2, \tag{4.25}$$

where σ_d is the standard deviation of e_d, the following equation results:

$$\mu_d^2 + \sigma_d^2 = \mu_e^4 + 6\mu_e^2\sigma_e^2 + 3\sigma_e^4. \tag{4.26}$$

Rearranging equation (4.22) gives

$$\mu_e^2 = \mu_d - \sigma_e^2. \tag{4.27}$$

Substituting equation (4.27) into (4.26) yields

$$(\mu_d - \sigma_e^2)^2 + 6(\mu_d - \sigma_e^2)\sigma_e^2 + 3\sigma_e^4 = \mu_d^2 + \sigma_d^2. \tag{4.28}$$

Expanding equation (4.28) and canceling like terms on both sides yields

$$-2\sigma_e^4 + 4\mu_d\sigma_e^2 - \sigma_d^2 = 0. \tag{4.29}$$

Solving for σ_e^2 in equation (4.29) yields

$$\sigma_e^2 = \mu_d \pm \sqrt{\mu_d^2 - \frac{\sigma_d^2}{2}}. \tag{4.30}$$

Substituting equation (4.30) into (4.27) gives

$$\mu_e^2 = \mu_d - \left(\mu_d \pm \sqrt{\mu_d^2 - \frac{\sigma_d^2}{2}}\right) = \mp\sqrt{\mu_d^2 - \frac{\sigma_d^2}{2}}. \tag{4.31}$$

Because $\mu_e^2 \geq 0$, ignoring the negative solution in (4.31) finally results in

$$\mu_e = \left(\mu_d^2 - \frac{\sigma_d^2}{2}\right)^{1/4}. \tag{4.32}$$

And substituting equation (4.32) into (4.27) yields

$$\sigma_e = \left(\mu_d - \sqrt{\mu_d^2 - \frac{\sigma_d^2}{2}}\right)^{1/2}. \tag{4.33}$$

Probability Analysis of r_2 and e_2

According to the derived mean and standard deviation of e_2, given by equations (4.32) and (4.33), the two reliability cases have been analyzed. Two probability plots are produced and presented in Figures 4.38 and 4.39.

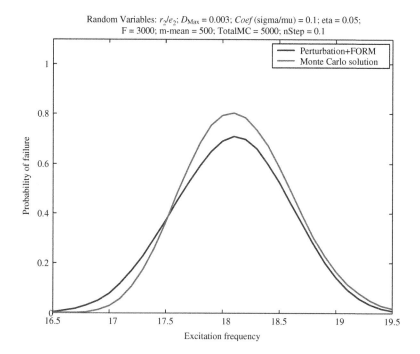

FIGURE 4.38   Reliability analysis of r_2 and e_2, $D_{\text{Max}} = 0.003$, $Coef = 0.1$.

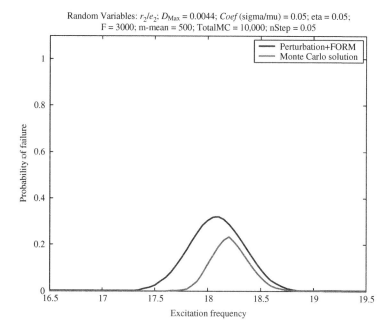

FIGURE 4.39   Reliability analysis of r_2 and e_2, $D_{\text{Max}} = 0.0044$, $Coef = 0.05$.

As seen in the two plots, though improved, the accuracy of the combined approach by employing e_2 is still not good enough to mimic the "exact" solution. Normal distribution analysis of e_2 is now considered.

Normal Distribution Analysis of e_2

The normal distribution analysis reveals that e_2 is still not normally distributed at the resonance, according to Figures 4.40–4.45, though being closer to a Gaussian variable in non-resonance cases.

Conclusion: e_2 is not a Gaussian variable either, so that the probability of failure predicted by employing r_2 and e_2 was still not accurate. Again, in order to apply the FORM method, a new parameter, which should be Gaussian at resonance, is needed to replace e_2. It is the only variable element in the definition of d_2, ω_2^2. The earlier approximation analysis of ω_2^2 was satisfied (Section 4.2.2), which suggests that it is a better candidate.

4.3.4 Solution 3: Variable ω_2^2 to Replace e_2

Normal distribution analysis of ω_2^2 was carried out before the reliability analysis. The analysis was directly performed at the resonance excitation frequency 18.11 Hz.

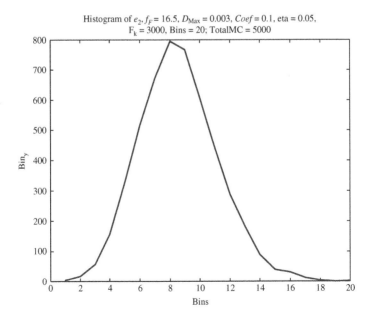

FIGURE 4.40 Histogram of e_2, $f_F = 16.5$, bins = 20.

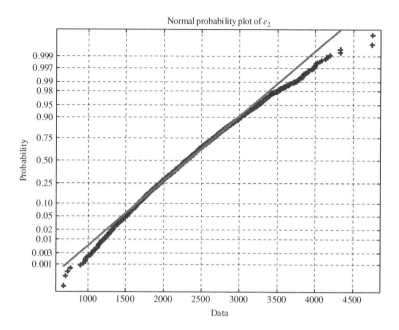

FIGURE 4.41   Normal probability of e_2, $f_F = 16.5$, bins = 20.

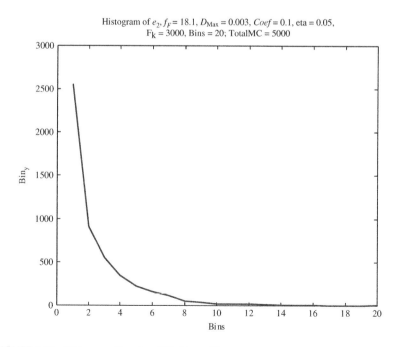

FIGURE 4.42   Histogram of e_2, $f_F = 18.1$, bins = 20.

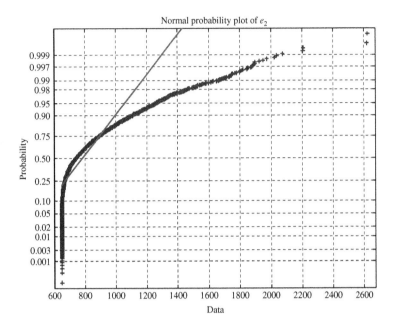

FIGURE 4.43   Normal probability of e_2, $f_F = 18.1$, bins = 20.

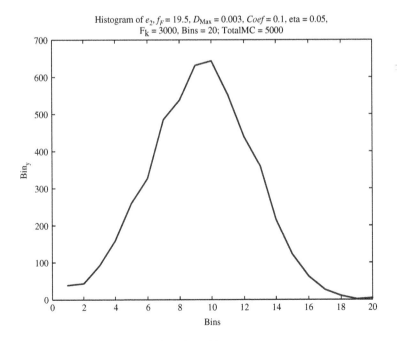

FIGURE 4.44   Histogram of e_2, $f_F = 19.5$, bins = 20.

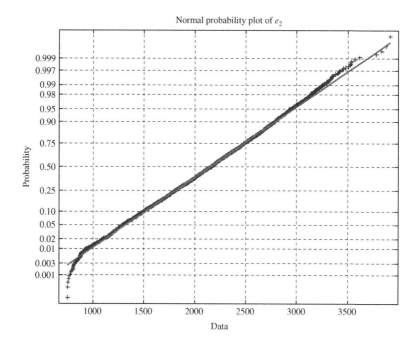

FIGURE 4.45   Normal probability of e_2, $f_F = 19.5$, bins $= 20$.

Normal Distribution Analysis of ω_2^2

The histogram plot and normal probability plot of ω_2^2 are presented in Figures 4.46 and 4.47 respectively. Even under the excitation frequency, ω_2^2 still constitutes a normal distribution, which proves that ω_2^2 is Gaussian.

Probability Analysis of r_2 and ω_2^2

The processes of finding the mean and standard deviation (covariance matrix) of r_2 remain the same as in equations (3.108) and (3.98) respectively. The mean of ω_2^2 can be determined directly by the eigen solution with the mean added masses. The standard deviation of ω_2^2 can be evaluated using equation (3.41). The reliability cases were re-run and the results are presented in Figures 4.48 and 4.49.

According to Figure 4.48, the most accurate prediction of the probability of failure delivered by the perturbation + FORM approach so far was achieved by employing r_2 and ω_2^2 as the random variables. However, there is still an obvious discrepancy between the results of the combined approach and that of the "exact" solution. Moreover, the discrepancy increases when the probability of failure becomes smaller, as shown in Figure 4.49.

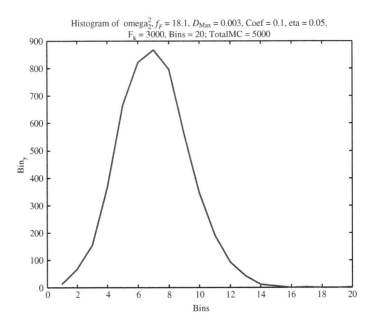

FIGURE 4.46 Histogram of ω_2^2, $f_F = 18.1$, bins = 20.

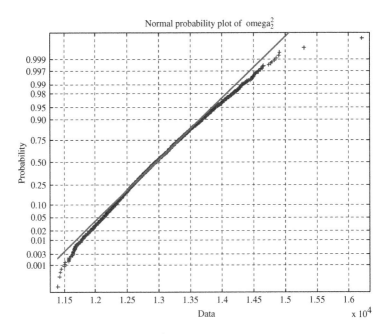

FIGURE 4.47 Normal probability of ω_2^2, $f_F = 18.1$, bins = 20.

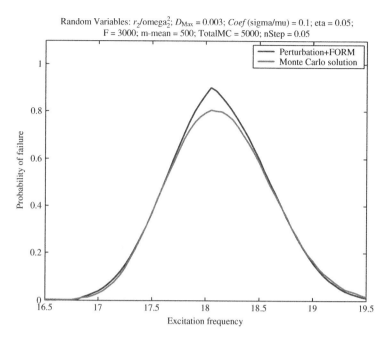

FIGURE 4.48   Reliability analysis of r_2 and ω_2^2, $D_{Max} = 0.003$, $Coef = 0.1$.

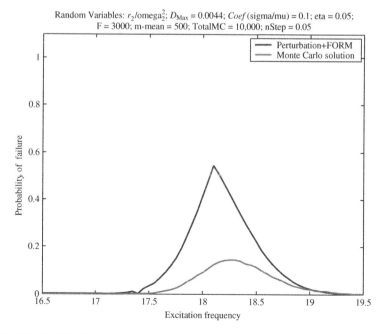

FIGURE 4.49   Reliability analysis of r_2 and ω_2^2, $D_{Max} = 0.0044$, $Coef = 0.05$.

Conclusion: Employing the two Gaussian variables r_2 and ω_2^2 in the perturbation + FORM approach still failed to accurately predict the probability of failure. Another condition for the FORM method to work accurately needs to be reviewed because of the introduction of a new parameter ω_2^2.

Problem of Contradiction of FORM Analysis

There is solid evidence to prove that r_2 and ω_2^2 are Gaussian random variables. However, other evidence also shows that the results of the FORM method involving these two random variables are not accurate. This leads to a reconsideration of the two conditions for the FORM method to work accurately. Since the Gaussian variables condition is satisfied, it leaves a question over the linearity of the safety margin constructed on these two variables.

To be consistent with the analysis presented in Section 3.2.2, two plots of safety margin contours on r_2 and ω_2^2, with two different values of $D_{\max}$, were produced and are presented in Figures 4.50 and 4.51.

According to the above figures, what was suspected was proved. The safety margin constructed using r_2 and ω_2^2 reverted to multi-connected and non-monotonic curves.

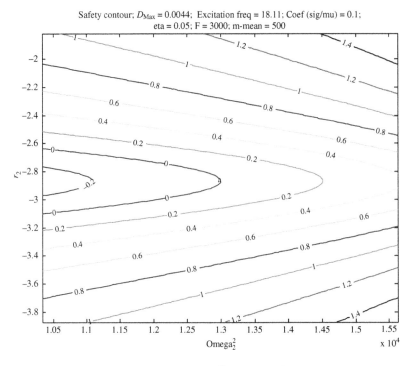

Safety contour; $D_{\text{Max}} = 0.0044$; Excitation freq = 18.11; Coef (sig/mu) = 0.1; eta = 0.05; F = 3000; m-mean = 500

FIGURE 4.50 Safety margin contours on r_2 and ω_2^2, $D_{\text{Max}} = 0.003$.

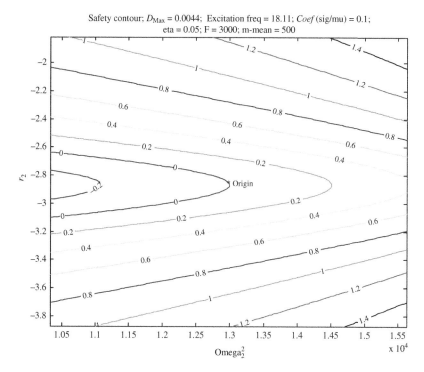

Safety contour; $D_{\text{Max}} = 0.0044$; Excitation freq = 18.11; $Coef$ (sig/mu) = 0.1; eta = 0.05; F = 3000; m-mean = 500

FIGURE 4.51   Safety margin contours on r_2 and ω_2^2, $D_{\text{Max}} = 0.0044$.

Conclusion: A contradiction problem was found in applying the FORM method. One of the random variables chosen initially was not Gaussian though the safety margin built was fairly linear. Replacing it with a Gaussian variable, however, increased the nonlinearity of the safety margin. The two conditions for the FORM method to be accurate, i.e. Gaussian variables and a linear failure surface, cannot be satisfied simultaneously with the selected pair of random variables. This contradiction problem results in a significant inaccuracy of the results delivered by the combined approach involving the FORM method.

4.4 SOLUTION 4: MONTE CARLO SIMULATION REPLACING FORM

The perturbation + FORM approach has been proved to be problematic as it does not deliver an accurate reliability prediction due to either the involvement of non-Gaussian random variables or nonlinear failure surfaces. However, the FORM method is only one of the available choices of probabilistic methods. Another one, the Monte Carlo simulation method, can be selected to replace FORM to evaluate the probability of failure by executing a number of MC simulations on the two Gaussian random variables r_2 and ω_2^2.

4.4.1 Perturbation + Monte Carlo Simulation on r_2 and ω_2^2

In order to apply the probabilistic Monte Carlo simulation method, it is required that the mean and standard deviation, as well as the covariance, of the variables are available. The mean and standard deviation (covariance matrix) of r_2 and ω_2^2 can be obtained by the method described in the last section. The covariance matrix of the parameters R_{jk} and Ω can be developed using a similar method to that introduced in Section 3.2.5.

According to the definitions given in Chapter 3, equations (3.79) and (3.34), the new combined parameter, denoted as $\tilde{T}$, can be written as

$$\tilde{T}_{jk} = \begin{pmatrix} \Omega \\ R_{jk} \end{pmatrix}_{2n \times 1} = \begin{pmatrix} \Delta\omega_1^2 & \Delta\omega_2^2 & \ldots & \Delta\omega_n^2 & \Delta r_{jk,1} & \Delta r_{jk,2} & \ldots & \Delta r_{jk,n} \end{pmatrix}^T_{2n \times 1}.$$

(4.34)

Substituting equation (3.37) and (3.91) into equation (4.34) yields

$$\tilde{T}_{jk} = \begin{pmatrix} A \\ F_k(S_j\tilde{Q}_k + S_k\tilde{Q}_j) \end{pmatrix}_{2n \times n^2} k - \begin{pmatrix} B \\ F_k(S_j\tilde{P}_k + S_j\tilde{W}_k + S_k\tilde{P}_j + S_k\tilde{W}_j) \end{pmatrix}_{2n \times n^2} m.$$

(4.35)

Letting

$$\tilde{U}_{jk} = \begin{pmatrix} A \\ F_k(S_j\tilde{Q}_k + S_k\tilde{Q}_j) \end{pmatrix}_{2n \times n^2}$$

(4.36)

and

$$\tilde{V}_{jk} = \begin{pmatrix} B \\ F_k(S_j\tilde{P}_k + S_j\tilde{W}_k + S_k\tilde{P}_j + S_k\tilde{W}_j) \end{pmatrix}_{2n \times n^2},$$

(4.37)

the combined parameter is described in matrix form as

$$\tilde{T}_{jk} = \tilde{U}_{jk}k - \tilde{V}_{jk}m.$$

(4.38)

Finally, the covariance matrix of $\tilde{T}$ can derived as

$$C_{\tilde{T}_{jk}} = E(\tilde{T}_{jk}\tilde{T}_{jk}^T) = \tilde{U}_{jk}C_k\tilde{U}_{jk}^T + \tilde{V}_{jk}C_m\tilde{V}_{jk}^T.$$

(4.39)

Therefore, the covariance of the two random variables, i.e. $Cov(r_2, \omega_2^2)$, can be extracted from the combined matrix $C_{\tilde{T}_{jk}}$ according to equation (4.39).

The working process of the proposed combined approach has now been updated and is illustrated in Figure 4.52.

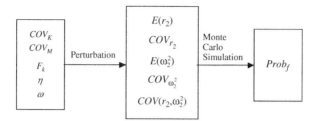

FIGURE 4.52   An updated working process of the combined approach.

4.4.2 Reliability Analysis of the Updated Combined Approach

The pseudo computer program for perturbation + MCS approach has been updated accordingly and is outlined below.

Perturbation + MCS

- Assemble the mass matrix $[M]$ with the mean added masses, m_1 and m_2.
- Call the eigen solver $eig(K,M)$ only once to obtain all the relevant modal properties, such as natural frequencies and mode shapes.
- Assemble the covariance matrix of $[M]$, C_m, and all other relevant perturbation matrices that do not involve the excitation frequency.
- Obtain the mean, standard deviation, and covariance values of r_2 and ω_2^2.
- Loop through excitation frequency (e.g. from 16.5 to 19.5 Hz), and for each excitation frequency:
  - Set *failure_number* to 0.
  - Loop through the number of the MC trials until TotalMC is reached.
    - Get a pair of realizations of the random variables of r_2 and ω_2^2 (using the known mean and standard deviation).
    - Evaluate the defined safety margin given in Table 4.5; if it is less than 0, which represents "failed", increase *failure_number* by 1.
  - End of Monte Carlo simulation loop.
  - Obtain the probability of failure under the current excitation frequency:

    Prob_F(currentFreq) = *failure_number*/TotalMC

- End of excitation frequency loop.

According to the updated program, the two designed cases were re-analyzed and the results are presented in Figures 4.53 and 4.54. The results are almost identical to the reference Monte Carlo FE solution.

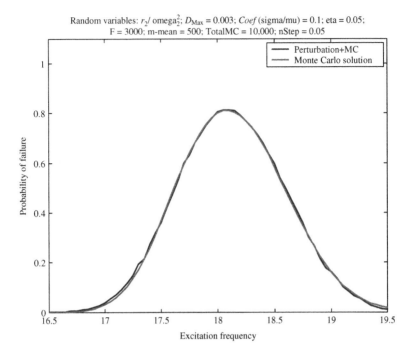

FIGURE 4.53 Reliability analysis of r_2 and ω_2^2 (Monte Carlo), $D_{Max} = 0.003$, $Coef = 0.1$.

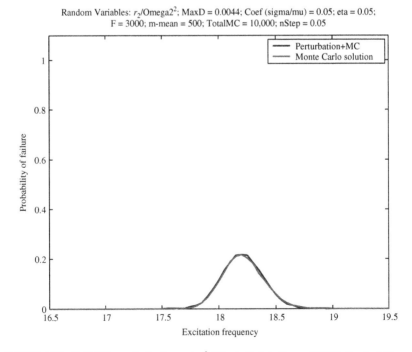

FIGURE 4.54 Reliability analysis of r_2 and ω_2^2 (Monte Carlo), $D_{Max} = 0.0044$, $Coef = 0.05$.

To test the combined approach on more cases, two other settings are introduced:

III. $D_{Max} = 0.0044$, $Coef = 0.1$, $\eta = 0.05$, a case of an increased variation compared to case II;
IV. $D_{Max} = 0.022$, $Coef = 0.05$, $\eta = 0.01$, a case of a smaller damping loss factor and D_{Max} is set on failure surface.[13]

The analysis of Cases III and IV were performed. Two results are presented by two plots in Figures 4.55 and 4.56 respectively.

Curves were matched precisely. It is observed that the reliability results predicted by the perturbation + MCS approach are consistently accurate in all four cases.

Conclusion: Perturbation + MCS approach is a correct solution to this dynamic problem. Further analysis including efficiency analysis needs to be carried out on a more complex system.

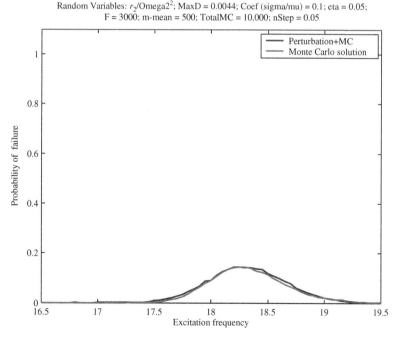

FIGURE 4.55   Reliability analysis of r_2 and ω_2^2 (Monte Carlo), $D_{Max} = 0.0044$, $Coef = 0.1$, $\eta = 0.05$.

---

13. According to the response analysis earlier (Figure 4.6), the resonance response is about 0.022 m at node 2 when $\eta = 0.01$.

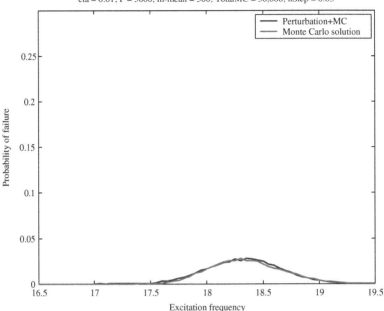

Random Variables: r_2/Omega2^2; MaxD = 0.022; Coef (sigma/mu) = 0.1; eta = 0.01; F = 3000; m-mean = 500; TotalMC = 50,000; nStep = 0.05

FIGURE 4.56 Reliability analysis of r_2 and ω_2^2 (Monte Carlo), $D_{\text{Max}} = 0.022$, $Coef = 0.1$, $\eta = 0.01$.

4.5 SUMMARY

The complete investigation process of the application of the combined approach to the 2D frame model is summarized in the chart presented in Figure 4.57.

The discoveries and solutions are summarized below:

1. According to Figures 4.5 and 4.6, the responses generated by the approximation modal model match very well to those by the "exact" FE solution at the resonance peak, particularly when the *damping loss factor* is low. This has proved that the simplified modal model, represented by equation (3.13) by neglecting all but the most significant terms at resonance in equation (3.10), is fairly accurate in approximating the resonance displacement.

2. The introduction of the new parameters, the pair r_2-d_2 in this case study, aims to overcome the nonlinearity problem of the failure surface when applying the FORM method. This aim has been achieved, according to Figures 4.9 and 4.10. The failure surfaces constructed on the pair r_2-d_2 have been transformed to fairly linear curves.

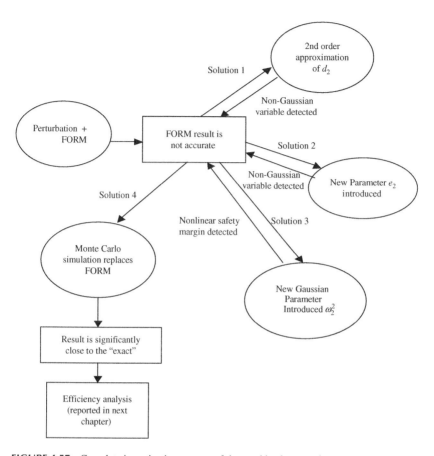

FIGURE 4.57   Complete investigation process of the combined approach.

3. The application of the perturbation + FORM approach failed, however, on using the pair r_2-d_2. In-depth studies have revealed later that the problem was because d_2 is not normally distributed, though r_2 is normally distributed (according to the conditions for the FORM method listed in Section 2.3). After further tests and analysis, ω_2^2 was finally chosen to replace d_2. The new pair r_2 and ω_2^2 are both Gaussian. However, problems were found as the failure surface constructed on the pair $r_2 - \omega_2^2$ was again nonlinear, as illustrated in Figures 4.50 and 4.51.

4. This contradictory problem was overcome by adopting the use of the Monte Carlo simulation method on the two Gaussian variables r_2 and ω_2^2. The accuracy of this application, i.e. the perturbation + MCS approach, was demonstrated for a variety of different cases, according to Figures 4.53−4.56.

TABLE 4.6 Summary of the Three Approaches

	Advantage	Disadvantage
Monte Carlo simulation on FE responses	**"Exact"**	**Computationally Expensive** Have to compute the inverse or to call the eigen solver many times
Perturbation approach + FORM	**Efficient** Only calls the eigen solver once	**Inaccurate** Particularly at resonance due to either nonlinear failure surface or non-Gaussian random variables
Perturbation approach + MCS	**Accurate and Efficient (to be tested)** Only calls the eigen solver once plus a set of Monte Carlo simulations only on the design variables	

The features of the three approaches adopted in this case study are compared in Table 4.6. The accuracy and efficiency of the final solution, i.e. perturbation approach + MCS, will be validated on a more complex system, i.e. a 3D helicopter dynamic system, in the next chapter.

Chapter 5

Application to a 3D Helicopter Model

With some amazing features such as rising vertically, hovering, and steering in any direction, helicopters are considered to be one of the greatest engineering achievements. To maintain such features, operation efforts are needed to provide lift, tilt, rotation, and torque compensation, which also means that helicopters are one of the most complex dynamic engineering systems [117]. The *harmonic excitation loads* generated by the spinning rotor, the *added random masses* due to the mounted equipment, e.g. vibration absorbers, and *a highly dynamic operation environment* make helicopters a perfect research model for the author's reliability approach.

This chapter presents an application of the combined approach, i.e. the perturbation approach followed by Monte Carlo simulations on defined parameter R_i and the selected natural frequency ω_i^2, to a 3D FE model of a helicopter fuselage.

5.1 BACKGROUND OF HELICOPTER VIBRATION CONTROL

The *dynamic rotor load* is a major vibration source in helicopter operations [118,119].[1] The amplitude of the load varies, usually increasing with the increase in forward speed [120,121]. This vibratory rotor load can be transmitted to different parts of the *fuselage*, causing vibration problems. Therefore, fuselage vibration reduction is an important design criterion in the helicopter industry, in responding to the constant demands for high speed, high performance, high system reliability, and limited maintenance costs [116,122,123].

There are two distinct vibration control methods, namely *passive control* and *active control* [116,121]. The passive control methods either utilize devices such as vibration absorbers, vibration isolators, and anti-resonance systems to suppress the vibration level at selected places, for example at the pilot's seat, or adopt structural modification and optimization to avoid the resonance. The active control methods involve real-time control technologies and can be classified into two categories: *rotor-based approaches* and

---

1. In addition to the rotor-induced vibration, boom tail vibration and grand resonance are two other vibration problems in helicopter dynamics [116,120].

Reliability Analysis of Dynamic Systems.
© 2013 Shanghai Jiao Tong University Press. Published by Elsevier Inc. All rights reserved. **119**

airframe-based approaches. Rotor-based approaches aim to reduce the vibratory loads that are transmitted through the hub to the airframe. Examples are higher harmonic control (HHC) and individual blade control (IBC), active flap control (AFC), and trailing edge flap (TEF) control [116,124,125]. Airframe-based approaches apply vibration reduction measures directly on the airframe, and include active control of structural response (ACSR), active vibration suppression (AVS), active vibration control (AVC), etc. [116]. While many researchers strongly believe that only the active control method has potential for future development [116], others claim that the conventional passive control methodologies are more reliable, more cost-effective, and simpler to implement [126].

Obviously, in order to provide an effective control solution, either passive, active or a combination, the key aspect is to *understand* and *measure* the vibrations and the corresponding responses [121,127]. Two primary objectives should be achieved thereafter: (1) to ensure that none of the predominant rotor excitation frequencies are close to any natural frequencies of the fuselage, i.e. to *avoid resonance*; and (2) to *reduce the dynamic responses* of the airframe under rotor-induced loads at the frequencies of interest [128].

It is expected that the proposed combined approach will provide a feasible reliability solution that will eventually contribute to the helicopter design analysis towards the above objectives.

5.2 A 3D HELICOPTER FE MODEL

According to the literature, *finite element* methods (software) are widely used to model different parts of helicopters, from rotor blades [129], fuselages [130,131], lifting surfaces [132], to under-floor planes, etc. [116]. Applications of the Monte Carlo simulation method [132] and response surface method [133] are also found in helicopter vibration research.

5.2.1 System Details

A finite element model of a 3D helicopter was constructed in FemLab. The geometry details of the model are presented in Figures 5.1 and 5.2 with the FE node index and edge index respectively. Since the structural data of a real helicopter is not available [134], this 3D model was a synthetic design in that all the settings were referred to and combined from different sources.[2] The geometry settings were iteratively modified so that the natural frequencies and mode shapes are similar to those of a real helicopter [134].

---

2. For example, the dimension details and geometry coordinates of the model were referred to different helicopter models available on the Internet. These were from commercial models such as MD600N, NH90, and EH101, to military ones such as US101, and research models [134]. So were the property settings and the location information of the added masses and other equipment, which were obtained from Refs [125,134,135].

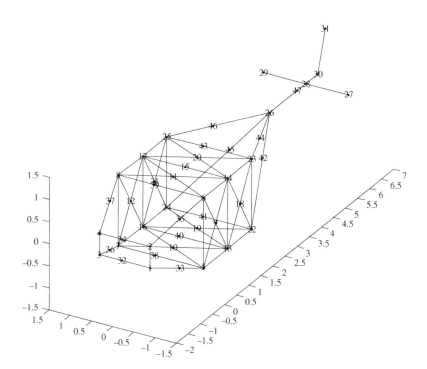

FIGURE 5.1   3D helicopter FE model with 47 nodes.

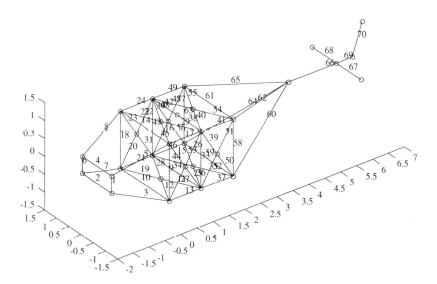

FIGURE 5.2   3D helicopter FE model with 70 elements.

TABLE 5.1 Dimensions of the Helicopter Model

Dimensions	Length (m)	Width (m)	Height (m)
Whole body	7	3	2
Fuselage	4	2.5	1.5

Based on the above considerations, the helicopter structure was defined to be 7 meters long, 3 meters wide, and 2 meters high. The length of the fuselage is 4 meters, the width is 2.5 meters, and the height is 1.5 meters (Table 5.1). The tailboom is 4.25 meters in length, having a horizontal stabilizer with a span (3 meters wide) attached to the end. The structure was modeled by 47 nodes and 70 edge elements. Since the helicopter is a free structure, all six DOFs, i.e. three directional displacements and three in-plane rotations, are allowed at each node, resulting in a total number of DOFs of 282.[3] The details of the nodal coordinates and element connectivity of the helicopter FE model are presented in Appendices VI and VII respectively.

Modern helicopter fuselages are built in forged, extruded sheet aluminum alloys held together using steel, titanium, and aluminum fasteners [135]. Accordingly, the material of the model was set as an aluminum alloy, having a density of 2800 kg/m^3, Young's modulus 7.0×10^{10} N/m^2, and Poisson's ratio 0.35. The whole body frame is made of an *FE beam structure*[4] with a cross-section of 0.224×10^{-3} m^2, while the *lift* section (around the fuselage to support hub and fuselage) and the *tail* section are thicker, 0.9×10^{-3} and 1.4923×10^{-3} m^2 respectively, offering stronger support. All the material properties and cross-section values (in standard units), with the corresponding edge numbers that they are applied to, are listed in Table 5.2.

During real helicopter operation, a number of devices and equipment may be loaded, which include for example engines, passive/active vibration control devices, and cargo. Some of them may be random in weight.

It is assumed that, in this 3D model, there are two engines (each weighing 180 kg), one tail gearbox (weighing 40 kg), two end-plates (each weighing 2 kg), and one passive vibration control device, i.e. a vibration absorber (weighing 18 kg). These devices and equipment are represented by added lumped masses on some selected nodes, i.e. they are attached to the structure at the appropriate

---

3. Again, it should be noted that this FE model is not an accurate helicopter model, which requires much finer meshes. It aims to provide a larger analysis model than the previous 2D model to demonstrate the approach used here.

4. The reason for employing a beam structure rather than a shell or plate is to simplify the complexity of the analysis of a Femlab model; also, this complies with the approach adopted in some research literature [134].

TABLE 5.2 Edge Element Settings (Properties) of the Helicopter Model

Properties/ Element Number	1−32, 35−37, 39−41, 44−65 (Body Frame)	33, 34, 38, 42, 43 (Lift Frame)	66−70 (Tail Frame)
A (m²)	0.224×10^{-3}	0.9×10^{-3}	1.4923×10^{-3}
Iyy (m⁴)	2.9419×10^{-8}	3.075×10^{-7}	1.6881×10^{-6}
Izz (m⁴)	2.9419×10^{-8}	3.075×10^{-7}	1.6881×10^{-6}
J (m⁴)	$2Izz$	$2Izz$	$2Izz$
Height_y (m)	0.03	0.05	0.1
Height_z (m)	0.03	0.05	0.1
E (N/m²)	7.0×10^{10}	7.0×10^{10}	7.0×10^{10}
Rho (kg/m³)	2800	2800	2800
Nu	0.35	0.35	0.35

TABLE 5.3 Details of the Added Masses

Added Masses	Node No.	Weight (kg)
Engine 1	23	180
Engine 2	25	180
Vibration absorber	10	18
End plate 1	27	2
End plate 2	29	2
Tail gear	31	40

nodes. It is assumed that the two engines are placed on top of the fuselage. Each of them is supported by one node, nodes 23 and 25 respectively. The tail gearbox is placed at node 31. The two end-plates are attached to the horizontal stabilizer at nodes 27 and 29 respectively. The absorber is placed near the pilot's seat, at node 10. The details of these settings are listed in Table 5.3.

5.2.2 Dynamic Characteristics of the Model

The first 20 *natural frequencies* of the system, with and without the added masses, are summarized in Table 5.4. These frequencies are closely spaced,

TABLE 5.4 Natural Frequencies of the Helicopter Model

Mode	Natural Frequencies (Original) (Hz)	Natural Frequencies (with Added Masses) (Hz)
1	3.94	1.54
2	4.12	1.92
3	5.24	3.03
4	14.91	9.40
5	17.43	11.10
6	21.88	12.28
7	22.39	14.02
8	22.66	21.04
9	23.07	21.70
10	23.89	22.02
11	26.31	22.32
12	26.52	23.56
13	27.44	24.33
14	28.22	25.27
15	28.67	26.19
16	29.51	26.40
17	31.84	26.93
18	36.74	27.06
19	40.45	30.06
20	41.22	30.58

which is the case for a real helicopter model [134]. Six mode shapes corresponding to the first six natural frequencies of the structure (with added masses) are presented in Figures 5.3–5.8. For example, the first mode corresponds to tailboom bending in the horizontal plane, while the second mode represents bending in the vertical plane and the third one corresponds to the torsion of the tailboom.

Depending on the number of rotor blades, the rotor excitation frequency of a real helicopter is normally between 10 and 20 Hz [136]. The excitation frequency of the harmonic rotor force in this application is therefore chosen as 11.1 Hz at mode 5 (see Table 5.4).

eigfreq(1)=1.54Line: total disp. (disp) Displacement: [x (u), y (v), z (w)]

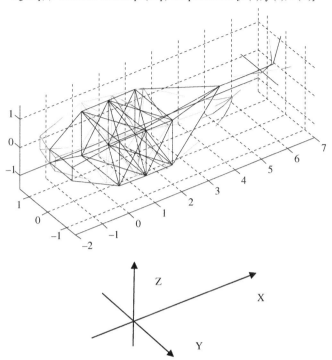

FIGURE 5.3   Mode shape 1 (tail yaw mode, bending in horizontal plane).

eigfreq(2)=1.92 Line: total disp. (disp) Displacement: [x (u), y (v), z (w)]

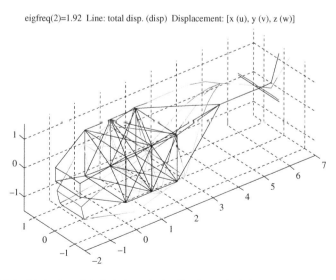

FIGURE 5.4   Mode shape 2 (tail pitch mode, bending in vertical plane).

eigfreq(3)=3.03 Line: total disp. (disp) Displacement: [x (u), y (v), z (w)]

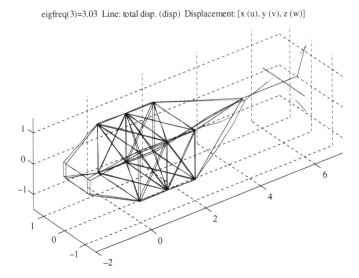

FIGURE 5.5   Mode shape 3 (tail roll mode, torsion of the tailboom).

eigfreq(4)=9.40 Line: total disp. (disp) Displacement: [x (u), y (v), z (w)]

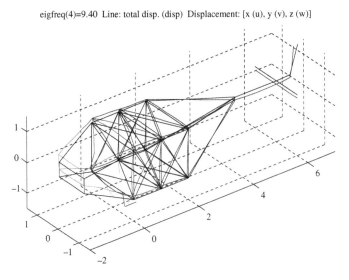

FIGURE 5.6   Mode shape 4 (body pitch mode + floor bending).

eigfreq(5)=11.10 Line: total disp. (disp) Displacement: [x (u), y (v), z (w)]

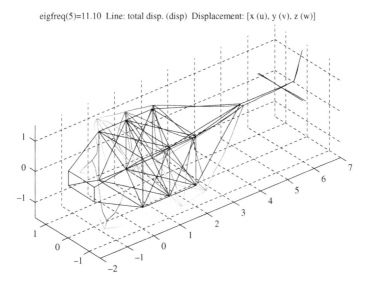

FIGURE 5.7   Mode shape 5 (body roll mode + torque).

eigfreq(6)=12.28 Line: total disp. (disp) Displacement: [x (u), y (v), z (w)]

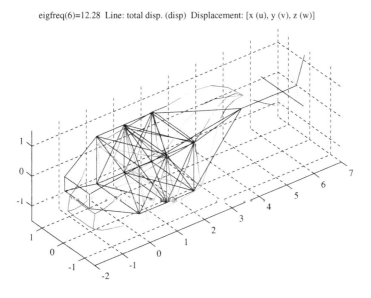

FIGURE 5.8   Mode shape 6 (second body pitch mode, bending in vertical plane).

5.3 RESPONSE ANALYSIS

To follow the practical requirement, the node for displacement/reliability analysis should be selected on the framework of the fuselage, near the pilot's seat. It was assumed therefore that it is the vicinity of node 9 where the pilot would be seated, and where there would be greatest concerns about the response in terms of safety and comfort. The displacement in the Y direction of node 9, i.e. lateral displacement, is chosen for the response analysis, which is reasonable for mode 5. Rotor force is placed at node 15 where the rotor is attached to the fuselage. The force amplitude is assumed to be 3000 N. To simplify the complexity of the dynamic analysis, it is further assumed that only Y direction rotor force at node 15 takes effect. The subscription of the defined parameter $r_{jk,r}$ was therefore determined as $j = 56$, $k = 62$, and $r = 5$.

By applying the response analysis approaches (1) and (3) designed in Chapter 4,[5] the responses at node 9 were analyzed under different excitation frequencies, with two defined damping loss factor settings, i.e. 0.05 and 0.01. The results are presented in Figures 5.9 and 5.10.

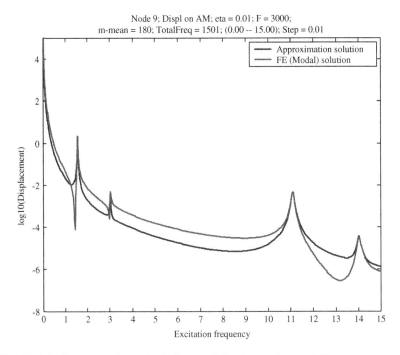

FIGURE 5.9 Response analysis of node 9 vs. excitation frequencies, $\eta = 0.01$.

---

5. As shown in Chapter 4, the results from approaches (1) and (2) are the same, thus only approach (1) is adopted here.

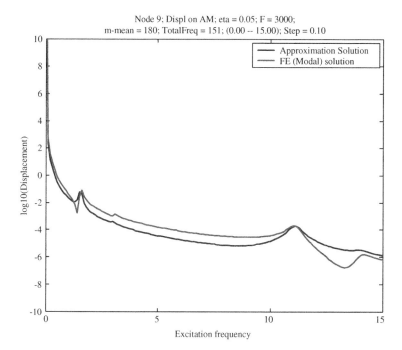

FIGURE 5.10 Response analysis of node 9 vs. excitation frequencies, $\eta = 0.05$.

As expected, the two curves generated by the approximation solution defined in equation (4.4) and the FE solution defined in equation (4.2) match almost perfectly at the resonance area. The match is closer when the damping loss factor η is lower, as shown in Figure 5.9 compared to Figure 5.10. This proves again that the proposed perturbation approach, an approximation solution, has accurately predicted the resonance displacement of the helicopter model under the designed excitation frequencies.

5.4 RELIABILITY ANALYSIS OF THE COMBINED APPROACH

5.4.1 Probability vs. Excitation Frequencies

According to the response analysis presented in Figures 5.9 and 5.10, the displacement of node 9 in the Y direction, under the resonance excitation frequency 11.1 Hz, is 0.06953 m when $\eta = 0.01$ and 0.01405 m when $\eta = 0.05$.[6]

---

6. To measure the damping loss factor of a real helicopter is a complex task, which depends on materials, dynamic frequency, temperatures, etc. Normally, it is reasonable to choose a factor value from 0.005 to 0.060 for a real helicopter when the excitation frequency is 15 Hz or under and temperature is less than 150°C [126,137,138].

Two cases are therefore set up for each η value. One is within the failure region and the other is on the failure surface. The details of the four cases are:

I. $D_{\text{Max}} = 0.05500$, $Coef = 0.1$, $\eta = 0.01$, within the failure region;
II. $D_{\text{Max}} = 0.06953$, $Coef = 0.1$, $\eta = 0.01$, on the failure surface;
III. $D_{\text{Max}} = 0.01390$, $Coef = 0.05$, $\eta = 0.05$, within the failure region;
IV. $D_{\text{Max}} = 0.01405$, $Coef = 0.05$, $\eta = 0.05$, on the failure surface,

where *Coef* is the coefficient of variation, defined as $Coef = \sigma/\mu$.

It is assumed that the two added masses representing the engine equipment are random, i.e. the two are mounted at nodes 23 and 25 respectively, and each with a mean value of 180 kg.[7] In practice, this randomness could affect all the relevant DOF elements in the helicopter mass matrix. To simplify the complexity and isolate the theoretical interests, it was further assumed that only the single element in the Y direction of nodes 23 and 25 in the mass matrix is random, while others are deterministic. This assumption applies to both the combined approach and the reference FE approach. The defined random parameters are obviously r_5 and ω_5^2. The program for the above cases was developed in Matlab, according to the procedure outlined in Section 4.4.2. The results are presented in Figures 5.11−5.14.

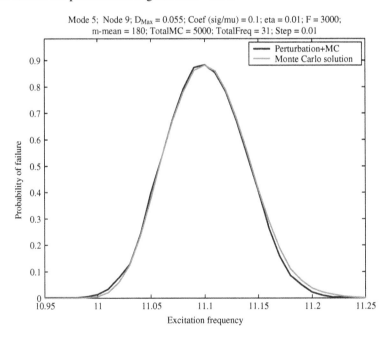

FIGURE 5.11   Reliability analysis of case I.

---

7. The standard deviation of the random masses depends on the variation coefficient, which is a parameter that varies for different cases.

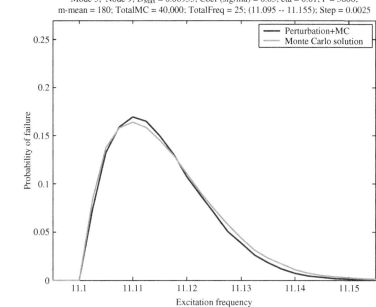

FIGURE 5.12   Reliability analysis of case II.

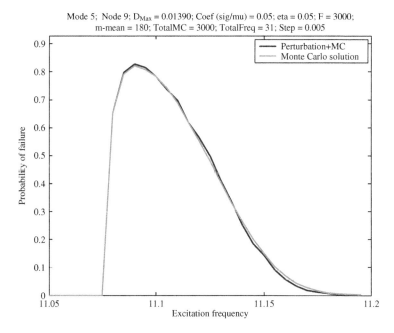

FIGURE 5.13   Reliability analysis of case III.

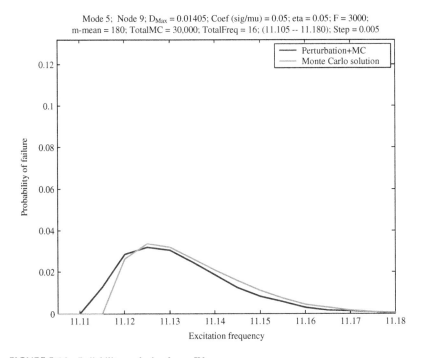

Mode 5; Node 9; $D_{Max} = 0.01405$; Coef (sig/mu) = 0.05; eta = 0.05; F = 3000;
m-mean = 180; TotalMC = 30,000; TotalFreq = 16; (11.105 -- 11.180); Step = 0.005

FIGURE 5.14   Reliability analysis of case IV.

All the results demonstrate an excellent match of the proposed combined approach with the FE "exact" solution, particularly when the damping loss factor η is low, which is consistent with the response analysis presented earlier. Also, when the probability of failure is higher, the match of the two curves is closer.

Figure 5.14 shows a small discrepancy between the two curves. This is due to a higher η value that was applied and that only two added mass random variables were involved. Nevertheless, given the low scale of the probability (less than 4%), the combined approach is fairly accurate in approximating the FE solution.

Moreover, non-symmetric curves were revealed, e.g. Figures 5.12 and 5.14, where resonance peak shifts away from the excitation frequency. This is also due to the theoretical assumption, i.e. that only a small part of the mass elements were involved to randomize the system.

5.4.2 Probability vs. Maximum Displacement and Variation Coefficient

Four more cases were suggested to observe the probability predicted by the combined approach on some other varying parameters, for example the

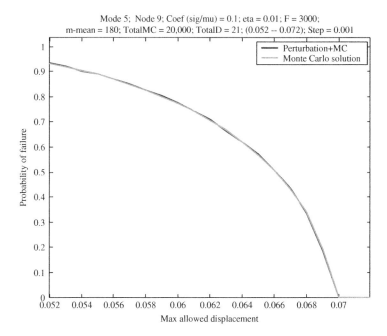

FIGURE 5.15 Reliability analysis of case V.

maximum allowed displacement (threshold value) and *variation coefficient*. The detailed settings of these new cases are:

  I. $Coef = 0.1$, $\eta = 0.01$, while D_{Max} varies;
 II. $Coef = 0.05$, $\eta = 0.05$, while D_{Max} varies;
III. $D_{Max} = 0.055$, $\eta = 0.01$, while $Coef$ varies;
 IV. $D_{Max} = 0.060$, $\eta = 0.01$, while $Coef$ varies.

The probabilities of failure of the above four cases are presented in Figures 5.15–5.18 respectively.

The results of cases V and VI, shown in Figures 5.15 and 5.16, demonstrated an almost identical match of the combined approach to the reference FE solution, which provides solid evidence that the combined approach is consistent and reliable in the cases of various D_{Max} given that the variation coefficient and the damping loss factor are fixed as 0.05–0.1 and 0.01–0.05 respectively.

The success of the approximation solution continues up to case VII, in which the damping loss factor is set to the low value 0.01 and the maximum allowed displacement is set to 0.055 (within the failure region). There is a small discrepancy at the tail of the curves where the variation coefficient became very high. In case VIII (in which the damping loss factor is set to the high value 0.05 and the maximum allowed displacement is set on the

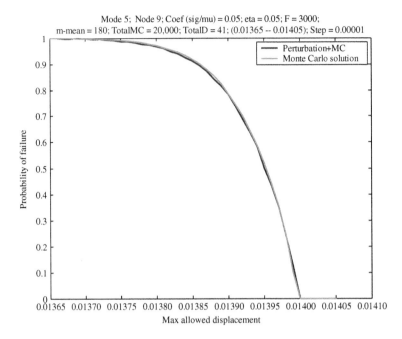

FIGURE 5.16 Reliability analysis of case VI.

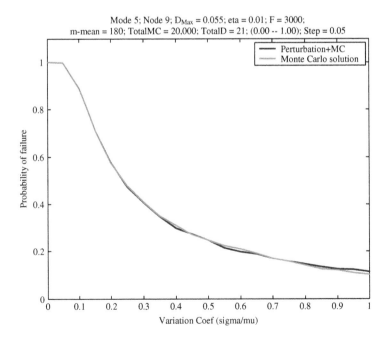

FIGURE 5.17 Reliability analysis of case VII.

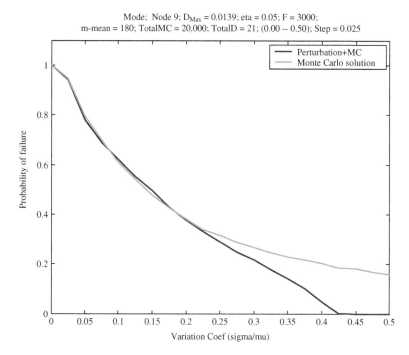

Mode; Node 9; D_{Max} = 0.0139; eta = 0.05; F = 3000;
m-mean = 180; TotalMC = 20,000; TotalD = 21; (0.00 -- 0.50); Step = 0.025

FIGURE 5.18 Reliability analysis of case VIII.

relevant failure surface), this discrepancy becomes larger when the variation coefficient is higher than 0.2, according to Figure 5.18. This implies that there are conditions for the combined approach to work well, i.e. a low system damping loss factor and relatively low variations of the random variables.

5.5 EFFICIENCY ANALYSIS

The responses produced by the combined approach and that by the FE solution were well matched at the resonance area. Accurate predictions of the probability of failure have been observed in all the cases with various settings. An *efficiency analysis* has also been conducted during the probability analysis. The total time consumed[8] by the two comparable approaches in the above eight cases are presented in Table 5.5.

According to the data in Table 5.5, a huge reduction of computing time is achieved using the combined approach. The total time consumed was reduced from days to seconds, without sacrificing the prediction accuracy.

---

8. The whole program was run on a PC with a 1.7 GHz processor and 512 MB RAM or equivalent.

TABLE 5.5 Time Consumed by the Two Approaches (hh:mm:ss)

Case Number	Direct Monte Carlo Simulation		Perturbation Approach + Monte Carlo Simulation	
	Total time (hh:mm:ss) (total number of units*/total number of MC trials)	Average time per unit (seconds)/per MC trial (seconds)	Total time (hh:mm:ss) (total number of units/total number of MC trials)	Average time per unit (seconds)/per MC trial (seconds)
I	11:38:38 (31/5000)	1352.19 (0.27)	00:00:22 (31/5000)	0.71 (0.000142)
II	46:17:21 (25/40,000)	4988.78 (0.25)	00:00:57 (25/40,000)	2.28 (0.000057)
III	06:44:41 (31/3000)	783.26 (0.26)	00:00:18 (31/3000)	0.58 (0.000194)
IV	155:28:17 (25/100,000)	22387.88 (0.22)	00:02:01 (25/100,000)	4.84 (0.000048)
V	41:02:39 (21/20,000)	7036.14 (0.35)	00:00:40 (21/20,000)	1.90 (0.000095)
VI	80:49:00 (41/20,000)	7096.10 (0.35)	00:01:05 (41/20,000)	1.59 (0.000079)
VII	38:19:17 (21/20,000)	6569.38 (0.33)	00:01:33 (21/20,000)	4.43 (0.000221)
VIII	41:02:41 (21/20,000)	7036.24 (0.35)	00:01:32 (21/20,000)	4.38 (0.000219)

*The number of units represents the number of frequencies, displacement values or the coefficients tested in each case.

TABLE 5.6 Analytical Estimates of the Sample Size of Monte Carlo Simulation

Case No.	Example P_f	$N \geq$
I	0.10	4000
II	0.01	40,000
III	0.16	2500
IV	0.005	80,000
V–VIII	0.2	20,000

The MC sample sizes of the direct Monte Carlo simulations and the perturbation + Monte Carlo simulations in each case were determined by two methods: an analytical estimate and a plot estimate, presented in Section 2.2. According to the analytical estimate (2.12), $N \geq (k/\varepsilon)^2(1/P)$, the sample size of the MCS of a target probability of failure can be approximately evaluated, assuming the error is less than 10% with 95% confidence ($k \approx 2$). These results are presented for each case in Table 5.6.

The great efficiency of the new approach makes it possible to perform a plot analysis suggested in Figure 2.1 in order to determine the Monte Carlo sample size N. The results of the first two cases are presented in Figures 5.19 and 5.20 respectively. Clearly, N was converged at 5000 in case I and 36,000 in case II.

Therefore, according to the results of the two methods, the final values of N set for the first two cases were 5000 and 40,000 respectively. The sample sizes, N, for other cases were determined accordingly and are listed in Table 5.5.

5.6 SUMMARY

It can be concluded therefore that the combined approach, i.e. perturbation approach + Monte Carlo simulation method, is a fairly accurate and extremely efficient solution for reliability analysis of a complex dynamic system, particularly when the damping loss factor of the system is low and the variation of the random variables is not too high, which was demonstrated in this 3D example.

So far, several assumptions have been made during the development of the combined approach. Accordingly, the conditions for the new approach to work accurately are summarized below:

1. In the development of the new modal model (Section 3.1) on which the perturbation approach was based, an approximation was made to neglect

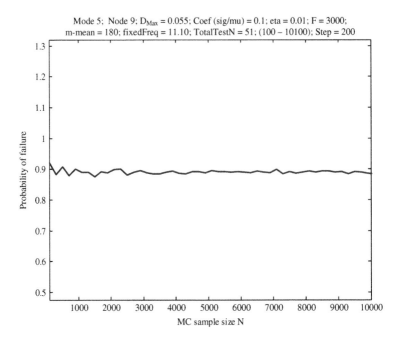

FIGURE 5.19   Plot analysis to determine the sample size of the new approach in case I.

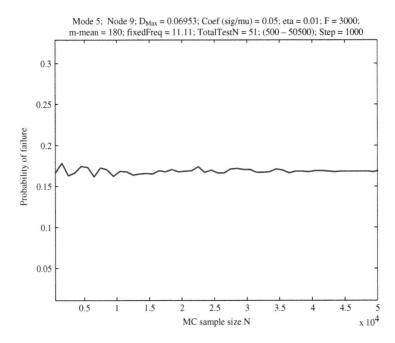

FIGURE 5.20   Plot analysis to determine the sample size of the new approach in case II.

the cross terms of the modal response. According to the response analysis presented in Figures 4.5, 4.6, 5.9, and 5.10, the responses were well approximated by the new approach at the resonance peaks. Thus, the relevant conditions are a *low modal overlap factor* and *resonance vicinity responses*. The definition of the low overlap factor was given in Chapter 3 as $M = \omega\eta n$. Clearly, the low damping loss will result in low modal overlap, which is consistent with the observations obtained in the two case studies. This implies that the author's approach should be capable of dealing with low-frequency dynamic problems.

2. The mean and covariance of r and ω^2 were estimated by perturbation analysis (Sections 3.2.1, 3.2.4, 3.2.6, and 4.4.1). Linear approximations were assumed in the derivation of the covariance. Therefore, the relevant condition is *low randomness*, which requires *low statistical overlap* and *small changes in system eigenvalues*.

3. Further, by applying Monte Carlo simulations on the new defined random parameters, it was assumed that r and ω^2 are Gaussian. This requires that either the *original variables are Gaussian themselves* and, through a *linear transformation*, or according to the *central limit theorem*, the new random parameters were influenced by (sum of) a number of arbitrary original random variables, regardless of their distribution types.

Chapter 6

Complete Combined Approach

Chapters 4 and 5 presented successful applications of the combined approach to a simple 2D frame structure and a complicated 3D helicopter structure. However, the approach is not yet complete, as it was based on the assumption that the covariance matrix of M, C_m, is known, while the stiffness matrix is assumed to be deterministic, i.e. $C_k = 0$. In other words, only the variation problem of added masses was treated, which randomizes the mass matrix M. In practice, however, the variation problem involves random structural properties, which influence the stiffness matrix K. Therefore, the covariance matrix of K, C_k, needs to be considered. This obviously requires derivation of C_k from the known statistical property information, e.g. the means and standard deviations of A, E, I, L, and ρ.

This chapter presents an efficient mapping approach using the response surface method (RSM), via a limited number of FE runs, to obtain an analytical relationship between the property random variables and the stiffness matrix, which can be described by the following equation:

$$K_{ij} = f(A, E, I, L, \rho, \text{etc.}), \tag{6.1}$$

where K_{ij} is an arbitrary element in the stiffness matrix K. Once such analytical relationships for all elements of K are determined, the required covariance matrix C_k can then be derived and assembled.[1]

The fundamentals of the RSM mapping technique are presented in Section 6.1. Full applications to both 2D frame and 3D helicopter models have been re-run and the results and analyses are reported in Sections 6.2 and 6.3.

6.1 RESPONSE SURFACE TECHNIQUES IN OBTAINING C_K

6.1.1 Direct RS Model Fitting of the Stiffness Matrix K

To fit equation (6.1), two RSM models were considered in the current investigation, i.e. types I and III (defined in Section 2.4.1), a linear and a quadratic model respectively.

---

1. A similar process can be applied to derive the covariance matrix of the mass matrix M. However, in this study, as presented in the previous chapters, a particularly simple application was adopted to enable random added masses to cover all the randomness of the mass matrix.

Reliability Analysis of Dynamic Systems.

It is assumed that there are in total N different random property variables in the system, denoted as $x_1, x_2, \ldots, x_N$. They will influence, more or less, the n by n stiffness matrix K, where n is the total number of DOFs in the system.

Type I Response Surface Model

In terms of the type I linear response surface (RS) model, elements in the matrix K can be described by

$$K_{11} = a_{0,11} + a_{1,11}x_1 + a_{2,11}x_2 + \ldots + a_{N,11}x_N$$
$$\ldots$$
$$K_{ij} = a_{0,ij} + a_{1,ij}x_1 + a_{2,ij}x_2 + \ldots + a_{N,ij}x_N \qquad (6.2)$$
$$\ldots$$
$$K_{nn} = a_{0,nn} + a_{1,nn}x_1 + a_{2,nn}x_2 + \ldots + a_{N,nn}x_N.$$

In general:

$$K_{ij} = a_{0,ij} + \sum_{r=1}^{N} a_{r,ij}x_r. \qquad (6.3)$$

According to Table 2.4, to solve for the $N + 1$ unknown coefficients a in any K_{ij}, using a saturated design method, the minimum number of FE runs is $N + 1$. There are n^2 elements in K to be fitted. Therefore, the total number of FE runs to fit the whole matrix K is $(N + 1)n^2$.

Type III Response Surface Model

For a second-order type III RS model, we have

$$K_{11} = a_{0,11} + a_{1,11}x_1 + \ldots + a_{N,11}x_N + a_{11,11}x_1^2 + \ldots + a_{NN,11}x_N^2$$
$$\ldots$$
$$K_{ij} = a_{0,ij} + a_{1,ij}x_1 + \ldots + a_{N,ij}x_N + a_{11,ij}x_1^2 + \ldots + a_{NN,ij}x_N^2 \qquad (6.4)$$
$$\ldots$$
$$K_{nn} = a_{0,nn} + a_{1,nn}x_1 + \ldots + a_{N,nn}x_N + a_{11,nn}x_1^2 + \ldots + a_{NN,nn}x_N^2.$$

In general:

$$K_{ij} = a_{0,ij} + \sum_{r=1}^{N} a_{r,ij}x_r + \sum_{r=1}^{N} a_{rr,ij}x_r^2. \qquad (6.5)$$

The minimum number of FE runs required for fitting this second-order RS model, using a saturated design method, is $2N + 1$. The total number of FE runs for the whole K matrix is therefore $(2N + 1)n^2$. If the *central composite design* (CCD) method[2] is employed, the number of FE calls required

---

2. CCD is an ideal and widely used sampling method for the second-order RS model. Details of the CCD method are given in Section 2.4.2.

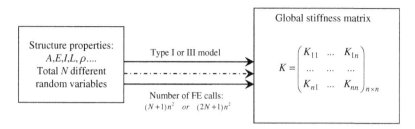

FIGURE 6.1   Fitting process of the global stiffness matrix using a situated design method.

for fitting each individual element is $N^2 + 2N + 1$, and that for the whole matrix will be $(N^2 + 2N + 1)n^2$.

Figure 6.1 demonstrates a process model of the RS technique of directly fitting the global stiffness matrix K. The process is straightforward and easy to implement. However, when N and n are large, it is difficult, and sometimes impossible, to apply such an approach. For example, for the given helicopter structure, the total number of DOFs is 282. Assuming that each element has only two different random variables, A and E for example, while others such as I, L, ρ are deterministic, the system would have a total of 140 random variables.[3] In summary, $N = 140$ and $n = 282$. The total number of the elements to fit in K is therefore $n^2 = 79{,}524$. The total number of FE calls required by a type I and type III RS model fitting are listed in Table 6.1.

Table 6.1 shows that the FE calls are huge for fitting either a type I or a type III model, even with the saturated design method. It is almost impossible to implement second-order RS fitting with the CCD method for this helicopter model as it requires more than 1.5 billion FE calls.

6.1.2 Alternative Fitting Approach

There is, however, a solution to the above problem. It fits all the local stiffness matrices instead and assembles all the local ones to a global stiffness matrix via matrix manipulation.

For example, for the two-node bar element shown in Figure 6.2, its stiffness matrix K in local coordinates xy is

$$K_{2 \times 2}^{\text{local}} = \frac{AE}{L} \begin{bmatrix} 1 & -1 \\ -1 & 1 \end{bmatrix}. \tag{6.6}$$

3. The total element number is 70 (edges).

TABLE 6.1 Total FE Calls Required by Two RS Models with Saturated and CCD Design Methods

	Saturated Design	Central Composite Design
First-order RS model	$(N+1)$ $n^2 = 11{,}212{,}884$	—
Second-order RS model	$(2N+1)$ $n^2 = 22{,}346{,}244$	$(N^2 + 2N + 1)n^2 = (140^2 + 2 \times 140 + 1) \bullet$ $282^2 = 1{,}581{,}016{,}644$

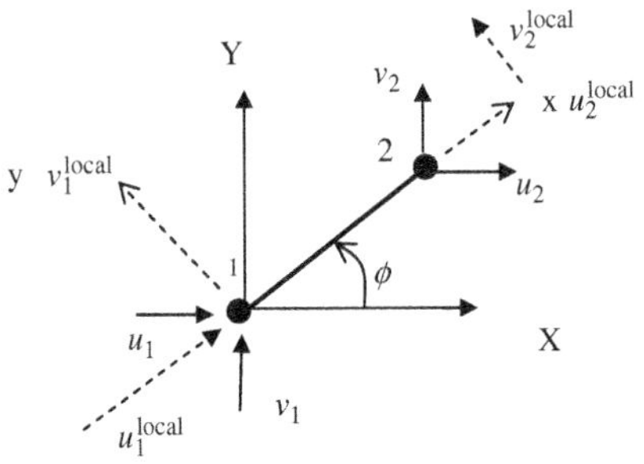

FIGURE 6.2   A two-node bar element with local and global coordinates.

The local displacement can be expressed in terms of the global coordinates in the following form:

$$\left\{ \begin{array}{c} u_1^{\text{local}} \\ v_1^{\text{local}} \end{array} \right\} = \begin{bmatrix} c & s & 0 & 0 \\ 0 & 0 & c & s \end{bmatrix} \left\{ \begin{array}{c} u_1 \\ v_1 \\ u_2 \\ v_2 \end{array} \right\}, \tag{6.7}$$

where $c = \cos \phi$, $s = \sin \phi$, and

$$T = \begin{bmatrix} c & s & 0 & 0 \\ 0 & 0 & c & s \end{bmatrix}$$

is called the *transformation matrix* [139].

The global stiffness matrix can be obtained in terms of $K_{2 \times 2}^{\text{local}}$ by the following manipulation:

$$K = T^T K_{2 \times 2}^{\text{local}} T = \frac{AE}{L} \begin{bmatrix} c^2 & cs & -c^2 & -cs \\ cs & s^2 & -cs & -s^2 \\ -c^2 & -cs & c^2 & c^2 \\ -cs & -s^2 & cs & s^2 \end{bmatrix}. \tag{6.8}$$

The above transformation operation provides an alternative but more efficient response surface fitting approach. When there are large numbers of property random variables (N) and DOFs (n), all the local stiffness matrices can be fitted separately and finally transformed and assembled together into the global stiffness matrix, according to the following equation:

$$K = \sum_{l=1}^{W} C_l^T T_l^T k_{m \times m}^l T_l C_l, \tag{6.9}$$

where W is the total number of elements in the structure, $k_{m \times m}^l$ is the local stiffness matrix with $m \times m$ local dimension (m is the number of the local DOFs), and T_l and C_l are the *transformation* and *connection* matrices corresponding to the element l [139]. T is responsible for transforming the local DOFs to the global DOFs, while C enables selection and placement of the relevant elements in the right position in terms of the global orientation. Finally, the global stiffness matrix K is a summation result of all the transformed and sorted local matrices.

Therefore, instead of working on a large global matrix, RS fitting can be undertaken on local stiffness matrices with a greatly reduced number of random variables. If N_L denotes the number of random variables contained in each element in the local stiffness matrix,[4] a type I *RS* model can be developed as

$$K_{11}^{local} = a_{0,11}^{local} + a_{1,11}^{local} x_1 + a_{2,11}^{local} x_2 + \ldots + a_{N_L,11}^{local} x_{N_L}$$
$$\ldots$$
$$K_{ij}^{local} = a_{0,ij}^{local} + a_{1,ij}^{local} x_1 + a_{2,ij}^{local} x_2 + \ldots + a_{N_L,ij}^{local} x_{N_L} \tag{6.10}$$
$$\ldots$$
$$K_{mm}^{local} = a_{0,mm}^{local} + a_{1,mm}^{local} x_1 + a_{2,mm}^{local} x_2 + \ldots + a_{N_L,mm}^{local} x_{N_L}.$$

In general:

$$K_{ij}^{local} = a_{0,ij}^{local} + \sum_{r=1}^{N_L} a_{r,ij}^{local} x_r. \tag{6.11}$$

This requires $N_L + 1$ FE calls by a saturated design method for one element. There are m^2 elements in this local stiffness matrix to be fitted. There is the same number of elements for the next local stiffness matrix until the total number of local matrices W described in equation (6.9) is obtained. Then the minimum total FE calls, using a saturated design method, is $(N_L + 1)m^2 W$.

---

4. It is assumed that each element has the same number of random variables N_L.

Similarly, for a type III RS model:

$$K_{11}^{\text{local}} = a_{0,11}^{\text{local}} + a_{1,11}^{\text{local}} x_1 + \ldots + a_{N_L,11}^{\text{local}} x_{N_L} + a_{11,11}^{\text{local}} x_1^2 + \ldots + a_{N_L N_L,11}^{\text{local}} x_{N_L}^2$$

$$\ldots$$

$$K_{ij}^{\text{local}} = a_{0,ij}^{\text{local}} + a_{1,ij}^{\text{local}} x_1 + \ldots + a_{N_L,ij}^{\text{local}} x_{N_L} + a_{11,ij}^{\text{local}} x_1^2 + \ldots + a_{N_L N_L,ij}^{\text{local}} x_{N_L}^2$$

$$\ldots$$

$$K_{mm}^{\text{local}} = a_{0,mm}^{\text{local}} + a_{1,mm}^{\text{local}} x_1 + \ldots + a_{N_L,mm}^{\text{local}} x_{N_L} + a_{11,mm}^{\text{local}} x_1^2 + \ldots + a_{N_L N_L,mm}^{\text{local}} x_{N_L}^2.$$

$$(6.12)$$

In general:

$$K_{ij}^{\text{local}} = a_{0,ij}^{\text{local}} + \sum_{r=1}^{N_L} a_{r,ij}^{\text{local}} x_r + \sum_{r=1}^{N_L} a_{rr,ij}^{\text{local}} x_r^2. \tag{6.13}$$

The minimum FE calls needed, using a saturated design method, is $(2N_L + 1)m^2 W$. For the CCD design method, $(N_L^2 + 2N_L + 1)m^2 W$ FE calls are required. Figure 6.3 illustrates this alternative fitting process.

Considering the helicopter example again; it has $N_L = 2$, $m = 12$, $W = 70$. The total number of elements for fitting a local stiffness matrix is only $12 \times 12 = 144$. The total number of FE calls for this alternative fitting is presented in Table 6.2.

Compared to the huge number of FE runs seen in Table 6.1, it is now possible to fit even the second-order RS model with the CCD design method using this local matrix fitting approach.

For later re-runs of reliability analysis of the two models, only the type I RS model with the *Koshal design method* (a *saturated design method*) was

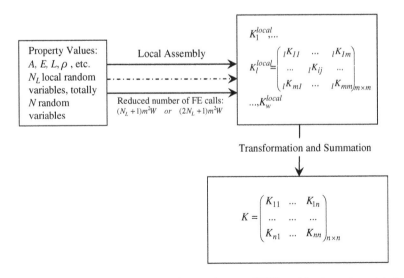

FIGURE 6.3 Alternative fitting process for type I or type III RS model using a situated design method.

TABLE 6.2 Total FE Calls Needed by Different RS Design Schemes for the Local Fit Approach

	Saturated Design	Central Composite Design (CCD)
First-order RS model	$(N_L + 1)m^2 W = 30{,}240$	–
Second-order RS model	$(2N_L + 1)$ $m^2 W = 50{,}400$	$(N_L^2 + 2N_L + 1)m^2 W = 90{,}720$

adopted and studied. The global matrix fitting process was selected. However, it was assumed that only a limited number of random variables were involved.

6.1.3 Analytical Approach to Obtain the Covariance Matrix of K

Once the linear analytical relationship between the original property random variables and the stiffness matrix elements has been established, the desired covariance matrix C_k can be derived as follows.

Assuming that there are N random variables,[5] denoted as $x_1, x_2, \ldots, x_N$, any two K matrix elements can be expressed as:

$$K_{ij} = a_{0,ij} + a_{1,ij}x_1 + a_{2,ij}x_2 + \ldots + a_{N,ij}x_N \tag{6.14}$$

and

$$K_{pg} = a_{0,pg} + a_{1,pg}x_1 + a_{2,pg}x_2 + \ldots + a_{N,pg}x_N. \tag{6.15}$$

The means of the two elements are therefore:

$$\overline{K}_{ij} = E(K_{ij}) = a_{0,ij} + a_{1,ij}\overline{x}_1 + a_{2,ij}\overline{x}_2 + \ldots + a_{N,ij}\overline{x}_N \tag{6.16}$$

and

$$\overline{K}_{pg} = E(K_{pg}) = a_{0,pg} + a_{1,pg}\overline{x}_1 + a_{2,pg}\overline{x}_2 + \ldots + a_{N,pg}\overline{x}_N. \tag{6.17}$$

By definition, the covariance of these two elements is

$$Cov(K_{ij}, K_{pq}) = E[(K_{ij} - \overline{K}_{ij})(K_{pg} - \overline{K}_{pg})]. \tag{6.18}$$

---

5. It is assumed that the mean and standard deviation of these random variables and the covariance between any pair of them are known.

Substituting equations (6.14)–(6.17) into equation (6.18) results in

$$
\begin{aligned}
Cov(K_{ij}, K_{pq}) &= E\left[\left(\sum_{r=1}^{n} a_{r,ij}(x_r - \bar{x}_r)\right)\left(\sum_{s=1}^{n} a_{s,pq}(x_s - \bar{x}_s)\right)\right] \\
&= \sum_{r=1}^{n}\sum_{s=1}^{n} a_{ij,r} a_{pq,s} Cov(x_r, x_s),
\end{aligned}
\tag{6.19}
$$

where $Cov(x_r, x_s)$ is the covariance of the property random variables x_r, x_s, which is a known value.

Finally, C_k, an n^2 by n^2 matrix, can be expressed as

$$
C_k = E(kk^T) = \begin{pmatrix}
Cov(K_{11}, K_{11}) & Cov(K_{11}, K_{21}) & \cdots & Cov(K_{11}, K_{nn}) \\
Cov(K_{21}, K_{11}) & Cov(K_{21}, K_{21}) & \cdots & Cov(K_{21}, K_{nn}) \\
\cdots & \cdots & Cov(K_{ij}, K_{pq}) & \\
Cov(K_{nn}, K_{11}) & Cov(K_{nn}, K_{21}) & \cdots & Cov(K_{nn}, K_{nn})
\end{pmatrix}_{n^2 \times n^2},
\tag{6.20}
$$

where $Cov(K_{ij}, K_{pq})$ is defined in equation (6.19).

Therefore, C_k can be obtained through a type I RS model fitting by solving for the coefficients a.

In particular, when $K_{ij} = K_{pq}$, i.e. for those diagonal elements in the covariance matrix C_k,

$$
Cov(K_{ij}, K_{ij}) = \sigma_{K_{ij}}^2 = \sum_{r=1}^{n} a_{ij,r}^2 \sigma_{x_r}^2 + \sum_{r=1}^{n}\sum_{\substack{s=1 \\ s \neq r}}^{n} a_{ij,r} a_{ij,s} Cov(x_r, x_s).
\tag{6.21}
$$

Further, when $Cov(x_r, x_s) = 0$ $(r \neq s)$, i.e. all the random variables are fully independent of each other, equations (6.19) and (6.21) can be simplified to

$$
Cov(K_{ij}, K_{pq}) = \sum_{r=1}^{n} a_{ij,r} a_{pq,r} \sigma_{x_r}^2
\tag{6.22}
$$

and

$$
Cov(K_{ij}, K_{ij}) = \sigma_{K_{ij}}^2 = \sum_{r=1}^{n} a_{ij,r}^2 \sigma_{x_r}^2.
\tag{6.23}
$$

As discussed in Chapter 2, an RS model is normally constructed using the so-called *coded variable b*, which is generally expressed as

$$
K_{ij} = b_{ij,0} + b_{ij,1}\frac{x_1 - \bar{x}_1}{\sigma_{x_1}} + \ldots + b_{ij,n}\frac{x_n - \bar{x}_n}{\sigma_{x_n}}.
\tag{6.24}
$$

Denoting the reciprocal of the coefficient of variation of x_r as $V_r = \bar{x}_r/\sigma_{x_r}$, and substituting it into equation (6.24), yields

$$K_{ij} = b_{ij,0} + \frac{b_{ij,1}}{\sigma_{x_1}} x_1 - V_1 b_{ij,1} + \ldots + \frac{b_{ij,n}}{\sigma_{x_n}} x_n - V_n b_{ij,n}$$

$$= \left(b_{ij,0} - \sum_{r=1}^{n} V_r b_{ij,r} \right) + \frac{b_{ij,1}}{\sigma_{x_1}} x_1 + \ldots + \frac{b_{ij,n}}{\sigma_{x_n}} x_n. \tag{6.25}$$

Comparing equation (6.25) to equation (6.3) shows that

$$a_{ij,0} = b_{ij,0} - \sum_{r=1}^{n} V_r b_{ij,r} \tag{6.26}$$

and

$$a_{ij,r} = \frac{b_{ij,r}}{\sigma_{x_r}}. \tag{6.27}$$

Substituting the above two equations into equations (6.21) and (6.22) respectively yields

$$Cov(K_{ij}, K_{pq}) = \sum_{r=1}^{n} b_{ij,r} b_{pq,r} \tag{6.28}$$

and

$$\sigma_{K_{ij}}^2 = \sum_{r=1}^{n} b_{ij,r}^2. \tag{6.29}$$

Therefore, by solving for the coefficients b in equation (6.24), C_k can be determined by equations (6.28) and (6.29) after fitting a linear response surface model.

6.1.4 Complete Combined Approach

The response surface method has been added to the perturbation + MCS approach to constitute a complete combined approach. Figure 6.4 presents the operational process of the approach.

The complete combined approach was applied to the 2D frame model and the 3D helicopter model. The results and discussion are presented in Sections 6.2 and 6.3 respectively.

6.2 COMPLETE APPLICATION TO 2D FRAME MODEL

6.2.1 Type I RS Model Fitting with Koshal Design

To simplify the randomness complexity of the 2D frame structure (Figure 4.1), it is assumed that there are two property variables, A_1, E_1 on element 1, that are random while others are deterministic.

According to the Koshal design scheme (a saturated design method) with two coded variables for fitting a type I model, three FE runs are therefore needed. The X matrix is defined as

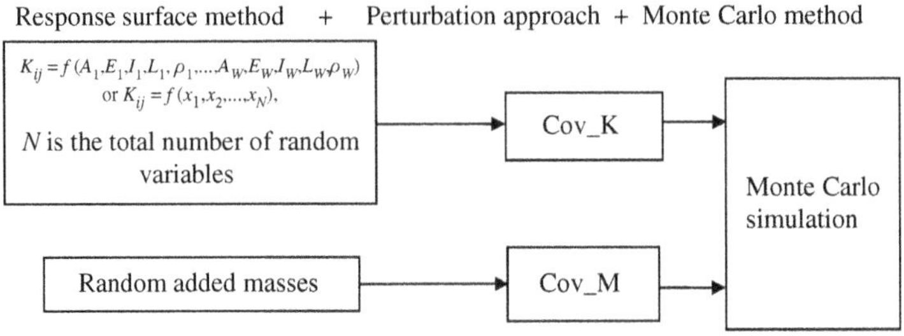

Response surface method   +   Perturbation approach + Monte Carlo method

$K_{ij} = f(A_1, E_1, I_1, L_1, \rho_1, \ldots, A_w, E_w, J_w, L_w, \rho_w)$
or $K_{ij} = f(x_1, x_2, \ldots, x_N)$.

N is the total number of random variables

→ Cov_K →

Random added masses ── → Cov_M →

Monte Carlo simulation

FIGURE 6.4   Process of the complete combined approach.

$$X = \begin{matrix} x_1 & x_2 \\ \begin{bmatrix} 1 & 0 & 0 \\ 1 & 1 & 0 \\ 1 & 0 & 1 \end{bmatrix} \end{matrix}. \tag{6.30}$$

The detailed FE run settings, in terms of original random variables, are

First FE call: $\overline{E}_1, \overline{A}_1$
Second FE call: $\overline{E}_1 + 1\sigma_{E_1}, \overline{A}_1$
Third FE call: $\overline{E}_1, \overline{A}_1 + 1\sigma_{A_1}$.

Each run (an FE solution) results in a global stiffness matrix K being assembled.[6] Rearranging all the 144 elements in a vector (column by column) and putting all the three vectors (after the three calls) together in a matrix generates a 3×144 matrix, denoted as K_{RS_Fit},

$$K_{RS\_Fit} = \begin{bmatrix} k_{11,1} & k_{21,1} & k_{31,1} & \cdots & k_{144144,1} \\ k_{11,2} & k_{21,2} & k_{31,2} & \cdots & k_{144144,2} \\ k_{11,3} & k_{21,3} & k_{31,3} & \cdots & k_{144144,3} \end{bmatrix} \begin{matrix} \text{First} \\ \text{Second} \\ \text{Third} \end{matrix}. \tag{6.31}$$

Since X is a square matrix, it easily yields the b coefficient matrix, B_{Coef}, as

$$B_{Coef} = X^{-1} K_{RS\_Fit} = \begin{bmatrix} b_{11,0} & b_{21,0} & b_{31,0} & \cdots & b_{144144,0} \\ b_{11,1} & b_{21,1} & b_{31,1} & \cdots & b_{144144,1} \\ b_{11,2} & b_{21,2} & b_{31,2} & \cdots & b_{144144,2} \end{bmatrix}. \tag{6.32}$$

Each column in the above matrix is a set of coefficients constituting one element in K in terms of coded variables, i.e.

$$K_{ij} = b_{ij,0} + b_{ij,1}x_1^c + \ldots + b_{ij,n}x_n^c, \tag{6.33}$$

where $x_i^c = (x_i - \overline{x}_i)/\sigma_{x_i}$ is the coded variable. Once the coefficient matrix B_{Coef} is obtained, C_k can be evaluated using equations (6.28) and (6.29).

---

6. K is a 12×12 dimensional matrix.

6.2.2 Complete Combined Approach

The computer program has been updated by adding the response surface method to derive the covariance matrix C_k, followed by the perturbation + MCS approach defined in Chapter 4, which is outlined below:

RSM + Perturbation + MCS

- Apply the RS model fitting technique with Koshal design method to obtain C_k according to the discussion in the last section.
- Assemble [K] and [M] with the mean values of A_1, E_1 and m_1, m_2.
- Call the eigen solver eig(K,M) only once to obtain all the relevant modal properties, such as natural frequencies and mode shapes.
- Assemble the covariance matrix of [M], C_m, and all other relevant perturbation matrices that do not involve the excitation frequency.
- Obtain the mean, standard deviation, and covariance values of r_2 and ω_2^2 using the C_k and C_m obtained.
- Loop by excitation frequency (e.g. from 16.5 to 19.5 Hz), and for each excitation frequency:
  - Set *failure_number* to 0.
  - Loop by the number of MC trials until TotalMC is reached:
    - Get a pair of realizations of the random variables of r_2 and ω_2^2 (using the known mean and standard deviation).
    - Evaluate the defined safety margin given in Table 4.5; if it is less than 0, which represents "failed", increases *failure_number* by 1.
  - End of Monte Carlo simulation loop.
  - Obtain the probability of failure under the current excitation frequency:
    - Prob_F(currentFreq) = *failure_number*/ TotalMC.
- End of excitation frequency loop.

The program for performing the Monte Carlo simulations on the FE solution needs to be updated as well, adding the simulation of A_1, E_1 random variables accordingly.

Monte Carlo Simulation of FE Solution ("Exact" Solution used as a Reference)

- Loop by excitation frequencies (e.g. from 16.5 to 19.5 Hz), and for each excitation frequency:
  - Set *failure_number* to 0.
  - Loop by the number of MC trials until TotalMC is reached:
    - Get a pair of realizations of the random variables of A_1, E_1 (using the known mean and standard deviation).
    - Assemble the stiffness matrix [K] with the two realizations.

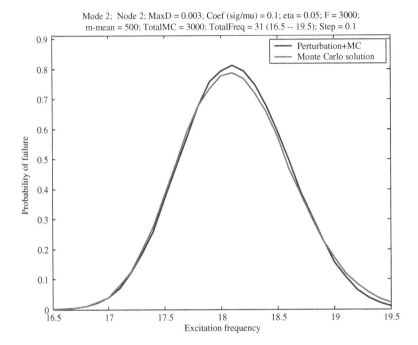

Mode 2; Node 2; MaxD = 0.003; Coef (sig/mu) = 0.1; eta = 0.05; F = 3000;
m-mean = 500; TotalMC = 3000; TotalFreq = 31 (16.5 -- 19.5); Step = 0.1

FIGURE 6.5   Reliability analysis of case I in Chapter 4 (re-run).

- • Get a pair of realizations of the random variables of m_1 and m_2 (using the known mean and standard deviation).
- • Assemble the mass matrix $[M]$ with the two added masses.
- • Process the FE solution, i.e. to solve equation (4.1), or use the modal solution, i.e. to solve the eigen problem eig(K,M) and evaluate equation (4.3) to obtain the displacement value on node 2, i.e. $\overline{X}_1$.
- • Evaluate the defined safety margin given in Table 4.5; if it is less than 0, which represents "failed", increase the *failure_number* by 1.
- • End of Monte Carlo simulation loop.
- • Obtain the probability of failure under the current excitation frequency:
  - • Prob_F(currentFreq) = *failure_number/*TotalMC.
- • End of excitation frequency loop.

The reliability analyses of cases I and II in Chapter 4 were re-run using the complete approach. The results are presented in Figures 6.5 and 6.6, and the relevant efficiency analysis is summarized in Table 6.3. Using the same method as presented in Chapter 5, the sample size (N) of MCS in each case was determined by the analytical estimate (but here $\varepsilon = 10\%$, $k \approx 2$) and the plot estimate.

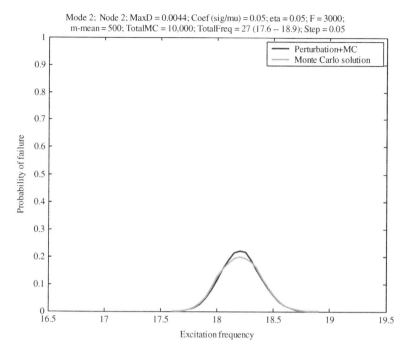

FIGURE 6.6   Reliability analysis of case II in Chapter 4 (re-run).

TABLE 6.3 Time Consumed by the Two Approaches for the 2D Model (hh:mm:ss)

Case Number (Chap. 4)	Direct Monte Carlo Simulation		Perturbation Approach + Monte Carlo Simulation	
	Total time (hh: mm:ss) (total number of units/ total number of MC trials)	Average time per unit (seconds)/per MC trial (seconds)	Total time (hh: mm:ss) (total number of units/ total number of MC trials)	Average time per unit (seconds)/per MC trial (seconds)
I	33:35:18 (31/ 3000)	3900.58 (1.30)	00:00:06 (31/ 3000)	0.18 (0.000061)
II	74:09:56 (27/ 10,000)	9888.74 (0.98)	00:00:16 (27/ 10,000)	0.61 (0.000061)

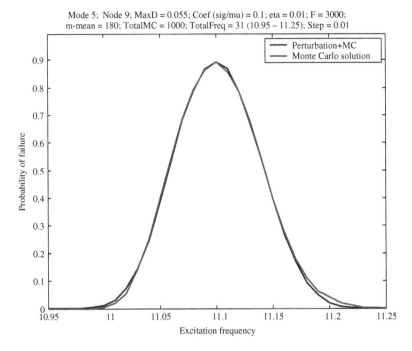

FIGURE 6.7   Reliability analysis of case I in Chapter 5 (re-run).

Discussion

Again, it has been observed that the applications of the complete approach with added RS method to the 2D model were very successful. According to the results presented in Figures 6.5 and 6.6 and Table 6.3, the accuracy and efficiency were guaranteed with an approximate C_k involvement, although small discrepancies were seen in both cases, which can be ignored (the error is less than 5%).

Compared to the analytical plots in Chapter 4 (Figures 4.53 and 4.54), little effect of the involvement of C_k regarding the probability results was found. This may be due to the boundary conditions of node 1 of the 2D model. However, in terms of the performance, it takes days to complete a large number of Monte Carlo simulations on FE for the assembling of both the stiffness and mass matrices under many excitation frequencies.[7] The new combined approach, in contrast, finishes the same analysis in just seconds (see Table 6.3)!

---

7. It took only minutes to run these cases without C_k involvement in Chapter 4.

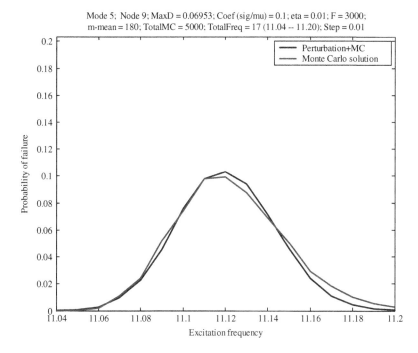

Mode 5; Node 9; MaxD = 0.06953; Coef (sig/mu) = 0.1; eta = 0.01; F = 3000;
m-mean = 180; TotalMC = 5000; TotalFreq = 17 (11.04 -- 11.20); Step = 0.01

FIGURE 6.8   Reliability analysis of case II in Chapter 5 (re-run).

6.3 COMPLETE APPLICATION TO 3D HELICOPTER MODEL

Similar to the previous example, to simplify the complexity issue, it was assumed that there are only two random property variables, A and E at element 26 of the helicopter model, which affect the stiffness of nodes 9 and 14. Therefore, a 12×12 stiffness matrix and 144×144 C_k were used instead of the whole corresponding 282×282 and $79,524 \times 79,524$ matrices.

The reliability analyses of cases I, II, III and IV of Chapter 5 were performed again according to the program described in the last section with involvement of C_k. The results are presented in Figures 6.7–6.10. The efficiency analysis is summarized in Table 6.4. Again, the sample size (N) of MCS in each case was determined by the analytical estimate ($\varepsilon = 10\%$, $k \approx 2$) and the plot estimate.

Discussion

In cases with a high probability of failure, i.e. those within the failure region (cases I and III), the predictions delivered by the complete combined approach are almost perfect. In case II (and case IV), the curve of the probability of failure produced by the complete approach is more symmetric than

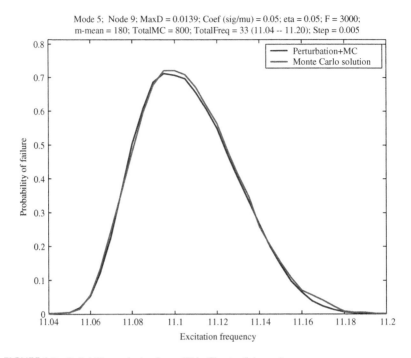

FIGURE 6.9   Reliability analysis of case III in Chapter 5 (re-run).

that presented in Chapter 5. This may be due to the addition of more random variables.

In cases with a low probability of failure (Figures 6.8 and 6.10), there are slight discrepancies between the two approaches. Nevertheless, given that the probability scale is low, the discrepancy is relatively small.

Overall, comparing to the reference Monte Carlo FE solution, the complete combined approach is fairly accurate in predicting the probability of failure of this complex 3D helicopter model.

The efficiency analysis is reported in Table 6.4. It takes the combined approach only minutes to finish a much larger number of Monte Carlo simulations. It should be noted that the time consumed in applying the combined approach includes the perturbation algorithm calculation time and the Monte Carlo simulation time. Involving operations on a number of relatively large-sized matrices, the algorithm calculation constitutes a dominant part of the total time consumed, which was about 6 minutes and 8 seconds for each case. Therefore, the total time (and the average time per unit) purely for executing Monte Carlo simulations in the combined approach for the four cases were 15 (0.48), 25 (1.47), 12 (0.36), and 23 (1.10) seconds respectively.

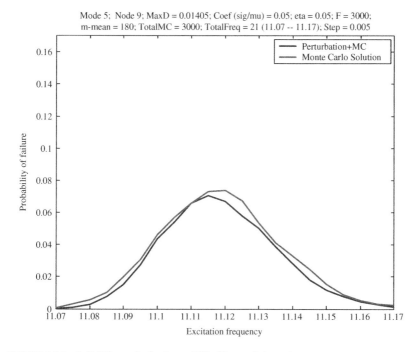

Mode 5; Node 9; MaxD = 0.01405; Coef (sig/mu) = 0.05; eta = 0.05; F = 3000; m-mean = 180; TotalMC = 3000; TotalFreq = 21 (11.07 -- 11.17); Step = 0.005

FIGURE 6.10   Reliability analysis of case IV in Chapter 5 (re-run).

The analysis has shown that the combined approach has great potential in running a much larger number of simulations, which is essential when analyzing a real reliability problem.

6.4 SUMMARY

To analyze the reliability of a very large and complex engineering structure with a number of random variables, the response surface method can be used as an efficient tool to approximate the unknown relationships that may otherwise require costly numerical solutions. In the previous case studies, comparing the RSM to the time consumed by the Monte Carlo finite element method, the application of the RSM in approximating the stiffness matrix significantly reduced the calculation time without excessively sacrificing accuracy.

The results of all six cases show that the saturated design method with the first-order response surface model fitting is efficient and sufficiently accurate. However, it should be pointed out that the random variables were assumed to be strictly limited, i.e. only two in both examples. If a large number of random variables were involved and the whole C_k was required, which would have dimensions of $n^2 \times n^2$ (the square of the total number of DOFs),

TABLE 6.4 Time Consumed by the Two Approaches for the 3D Model (hh:mm:ss)

Case Number (Chap. 5)	Direct Monte Carlo Simulation		Perturbation Approach + Monte Carlo Simulation	
	Total time (hh:mm:ss) (total number of units*/total number of MC trials)	Average time per unit (seconds)/per MC trial (seconds)	Total time (hh:mm:ss) (total number of units/total number of MC trials)	Average time per unit (seconds)/per MC trial (seconds)
I	44:34:21 (31/ 1000)	5176.16 (5.17)	00:06:23 (31/ 1000)	12.35 (0.01235)
II	87:37:41 (17/ 5000)	18,556.53 (3.71)	00:06:33 (17/ 5000)	23.12 (0.00462)
III	26:36:45 (33/ 800)	2903.18 (3.62)	00:06:20 (33/ 800)	11.51 (0.01439)
IV	78:17:44 (21/ 3000)	13,422.10 (4.47)	00:06:31 (21/ 3000)	18.62 (0.00621)

*The number of units represents the number of frequencies, displacement values or coefficients tested in each case.

the screening test, an RS technique that reduces the volume of the random variables by eliminating those that are "unimportant" [44], would become necessary. Moreover, second-order RS model fitting is definitely more accurate than its first-order counterpart, and after a screening test (with a reduced number of variables), it may be a better choice.

Conclusions will be drawn in Chapter 7. Some remaining issues will be discussed and enhancements of this combined method will be suggested.

Chapter 7

Conclusions and Future Work

7.1 ACHIEVEMENTS AND CONCLUSIONS

With the help of modern powerful computing capabilities, probabilistic approaches have been widely implemented in reliability analysis of engineering systems. The first-order reliability method (FORM), the Monte Carlo simulation (MCS) method, and the response surface method (RSM) are examples of these approaches. They are applied, often in combination with the finite element (FE) method, to quantitatively assess the probability of failure of engineering structures. However, there are either practical or theoretical limitations in applying these methods, particularly where dynamic systems are concerned.

Chapter 2 reviewed these problems in detail. Chapter 3 presented a combined approach as a solution to the problems. The proposed approach consists of a *perturbation approach* and one of the probabilistic methods.

The core techniques of the perturbation approach include:

1. A set of new random parameters that are defined on an approximation model of dynamics modal model.
2. A set of modal perturbation algorithms that are developed to obtain the statistical information of the new random parameters from known information of the original random variables.

However, when the FORM method was initially chosen to work with the perturbation approach for analyzing a 2D frame structure with random added masses, the probability of failure delivered by the combined approach was significantly inaccurate. In-depth analysis revealed that this was due to the fact that the conditions of the FORM method were breached, i.e. either a non-Gaussian random variable was involved or a nonlinear failure surface formed. Also, the two problems cannot be solved simultaneously. The Monte Carlo simulation method was then proposed to replace the problematic FORM method. Both accuracy and efficiency were achieved by this updated combined approach when it was applied to the 2D frame structure, as well as to a 3D helicopter model. The response surface method was implemented to derive the required covariance information of the stiffness matrix from that of the original property random variables, followed by a successful combined approach, which together offer an integrated and complete solution for reliability analysis of dynamic systems. Figure 7.1 illustrates a model of the combined approach.

The following conclusions can be drawn based on the successful applications of the combined approach presented in Chapters 4−6.

Firstly, the combined approach is technically accurate. All the analysis cases performed have shown good matches between the combined approach and the reference "exact" FE solution, particularly when the damping loss factor is low, where the two comparable results are almost identical. The precision has been kept consistently in different models as well as with different system settings.

Secondly, the combined approach is extremely efficient. The performance of the reference FE solution for analyzing the cases presented in Chapters 5 and 6 was measured in *days*, while that of the combined approach was in *minutes*. The huge reduction of computing time makes it possible to execute a larger number of simulations involving more random variables and complexities, which is important in engineering practice.

Thirdly, the combined approach has potential. As discussed in Chapter 1, there is a trend nowadays to combine the probabilistic reliability methods and possibilistic methods together. The modal-based approximation model and the perturbation algorithm can be extended for this demand. It is worth considering the development of a new set of algorithms to apply nonprobabilistic methods.

Due to the approximations made during the development of the approach, such as neglecting the cross terms in modal responses and linear approximations in the derivations of the statistical moments of the defined parameters, there are several **conditions for the success of the new approach**. These conditions are listed in Table 7.1.

7.2 FUTURE WORK

Three major investigations are recommended that will be very useful for future applications of the combined approach in both the academic and industrial fields.

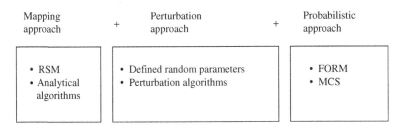

FIGURE 7.1 The process model of the combined approach.

TABLE 7.1 Conditions for the Success of the New Approach

Conditions
1 Low modal overlap factor
2 Responses near resonance
3 Low statistical overlap and small changes in eigenvalues
4 Gaussian distribution of the original variables or involvement of a number of arbitrary original random variables

---

7.2.1 Application of an Enhanced FORM Method

Chapter 4 presents an unsuccessful application of the FORM method to the 2D frame structure. The problem was due to the fact that either the defined parameters are not Gaussian or the failure surface, constructed by defined Gaussian variables, is highly nonlinear. The final solution adopted in the present work is to use the Monte Carlo simulation method to replace the FORM method. However, there may be an alternative solution that can still make use of the FORM method, but in an enhanced version.

According to the literature, the non-Gaussian variable problem in the FORM method has been addressed by some researchers. There were two major techniques developed in 1978 and 1981 respectively, namely the *Rackwitz−Fiessler (R-L) method* [23] and the *advanced first-order reliability method (AFORM)* [24].

The Rackwitz−Fiessler (R-L) method, named after the two researchers, is also referred to as the "equivalent normal" (EN) method. Its aim, through the incorporation of the available probability density function information (presumably non-Gaussian), is to develop a better Gaussian approximation to the true probability density functions in the area of interest within the design space, for example in the region of MPP [23], the *most probable point*. Figure 7.2 presents a graphical depiction of the approximation, given that $x^{\cdot}$ is the MPP.

Therefore, the R-F technique is to force the two density functions to share similar statistical properties in the area of interest, i.e. both density functions are *equivalent* at the MPP. Given a known cumulative density function (cdf) and the associated probability density function (pdf), denoted as $F(x)$ and $F(x^*)$ respectively, it is now desired to find the mean (μ) and standard deviation (σ) of an "equivalent" Gaussian density function such that the cdf and the pdf should satisfy:

$$F(x^*) = \Phi\left(\frac{x^* - \mu}{\sigma}\right) \tag{7.1}$$

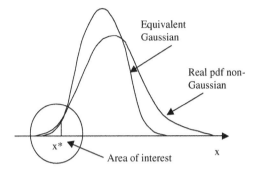

FIGURE 7.2   Equivalent normal pdf approximation.

and

$$f(x^*) = \frac{1}{\sigma}\phi\left(\frac{x^* - \mu}{\sigma}\right), \tag{7.2}$$

where $\Phi(\ )$ and $\phi(\ )$ are the standard normal cdf and pdf respectively.

Solving the two equations results in

$$\mu = x^* - \Phi^{-1}[F(x^*)]\sigma \tag{7.3}$$

and

$$\sigma = \frac{\phi\{\Phi^{-1}[F(x^*)]\}}{f(x^*)}. \tag{7.4}$$

These two moments can then be substituted into the moments used during the H-L FORM iterations described in Chapter 2, which will finally deliver the desired safety index.

The advanced first-order reliability method (AFORM) was originally proposed by Hohenbichler and Rackwitz, hereafter also referred to as the H-R method. In addition to incorporating the normalization procedure of the H-L technique and the pdf approximation technique of the R-F method, AFORM also addresses the dependency problem of the random variables. By employing the Rosenblatt transformation method, AFORM aims to transform the non-Gaussian, dependent random variables into a set of independent, standardized Gaussian random variables.

Therefore, the approximation technique from the R-F method and the transformation technique from AFORM can be introduced to solve the non-Gaussian problem of the defined parameter d_2 presented in Chapter 4. By using the equivalent and dependent Gaussian approximation in the area of the interest, it can then deliver a better safety index and a more accurate result, which would eventually enable the FORM method to still be a choice among the probabilistic methods for the combined approach.

7.2.2 Further Simplification of Perturbation/Analytical Algorithms

The extreme efficiency of the combined approach has been observed consistently during all the application cases. However, in the present work, an assumption was made that the randomness was only treated on some selected elements in the mass and stiffness matrices. If all the elements were concerned in a large and complex system with a number of DOFs (n), the computational cost would be prohibitively high, due to the perturbation algorithms involving operations on large-sized matrices with dimension of n^2. Therefore, simplification of the perturbation algorithms enabling them to accommodate more randomness should be considered, which will greatly increase the efficiency and practicality of the combined approach. For example, an obvious method is to make use of the symmetric feature of the covariance matrices to instantly reduce the size by half.

Some useful techniques were described but not implemented in this study, for instance the use of the type III response surface model for a more accurate fitting of the covariance matrix of the stiffness matrix and the efficient local assembling approach, described in Sections 6.1.1 and 6.1.2 respectively. In order to increase the accuracy and efficiency, particularly for large complex engineering systems, it would be very useful to test and implement these techniques in the future case studies.

7.2.3 Development for Non-Probabilistic Methods

As discussed in Chapter 1, though being still at a preliminary stage, the combination of the probabilistic and non-probabilistic methods has drawn attention from researchers all over the world. Development of a feasible combination technique could be the next major step in reliability analysis practice.

It has been proved that the approximation model developed in the present work, represented by equation (3.16), is accurate in approximating the resonance responses. Therefore, there is potential for the model to be used as a basis for developing approaches to apply non-probabilistic methods such as interval theory or fuzzy theory.

Transforming Random Variables from Correlated to Uncorrelated

Assume that μ_x, σ_x, and C_x represent a vector of mean values, a vector of standard deviations, and covariance matrix respectively, of a set of dependent random variables x, denoted as x_1, x_2, ..., x_n. It is desired to obtain μ_y, σ_y, and C_y, where the new C_y is a diagonal matrix, i.e. off-diagonal elements are all zero.

In terms of two variables, μ_x, σ_x, and C_x will be:

$$\mu_x = \left\{ \begin{matrix} \bar{x}_1 \\ \bar{x}_2 \end{matrix} \right\}, \quad \sigma_x = \left\{ \begin{matrix} \sigma_{x_1} \\ \sigma_{x_2} \end{matrix} \right\}, \quad \text{and} \quad C_x = \begin{pmatrix} \sigma_{x_1}^2 & \rho\bar{x}_1\bar{x}_2 \\ \rho\bar{x}_1\bar{x}_2 & \sigma_{x_2}^2 \end{pmatrix},$$

where ρ is the correlation coefficient, defined as $\rho = Cov(x_1, x_2)/\sigma_{x_1}\sigma_{x_2}$, and $\rho\bar{x}_1\bar{x}_2 \neq 0$.

It is required to get

$$\mu_y = \left\{ \begin{matrix} \bar{y}_1 \\ \bar{y}_2 \end{matrix} \right\}, \quad \sigma_y = \left\{ \begin{matrix} \sigma_{y_1} \\ \sigma_{y_2} \end{matrix} \right\}, \quad \text{and} \quad C_y = \begin{pmatrix} \sigma_{y_1}^2 & 0 \\ 0 & \sigma_{y_2}^2 \end{pmatrix}.$$

To obtain the uncorrelated variables the eigen problem of C_x needs to be solved to get the eigenvalues and eigenvectors, i.e. $\text{eig}[p,e] = \text{eig}(C_x)$, where e is a diagonal matrix of eigenvalues and p is a full matrix whose columns are the corresponding eigenvectors.

Then the required uncorrelated variables can be obtained by

$$\mu_y = p\mu_x \tag{AI.1}$$

and

$$\sigma_y = \text{diag}(e)^{1/2}. \tag{AI.2}$$

Analytical Solution of HL Safety Index

β should be easily obtained through an iteration process satisfying the conditions given in equation (2.18), so too the design point.

By using the technique of Lagrange multipliers, Shinozuka [108] found that the minimum distance represented by the HL safety index is

$$\beta_{HL} = \frac{-\nabla^{*T} z^*}{(\nabla^{*T}\nabla^*)^{1/2}} = \frac{-\sum_{i=1}^{n} z_i^* \left.\frac{\partial G(z)}{\partial z_i}\right|_{z=z^*}}{\sqrt{\sum_{i=1}^{n}\left(\frac{\partial G(z)}{\partial z_i}\right)^2\Bigg|_{z=z^*}}}, \tag{AII.1}$$

where the gradient is defined as

$$\nabla^{*T} = \left(\frac{\partial G}{\partial z_i}, \ \ldots, \ \frac{\partial G}{\partial z_n}\right)\Bigg|_{z=z^*}.$$

An excellent first-order interpretation of this expression (AII.1) was given by Ang and Tang [109], which is demonstrated as follows.

In general, for a nonlinear limit state function $G(x)$ involving many random variables, the problem can be solved by taking the Taylor series expansion of $G(x)$ at the MPP, i.e. $x^* = (x_1^*, \ldots, x_n^*)$. This gives

$$G(x) = G(x_1, \ldots, x_n) = G(x)\big|_{x=x^*} + \sum_{i=1}^{n}\frac{\partial G(x)}{\partial x_i}\Bigg|_{x=x^*}(x_i - x_i^*)$$

$$+ \frac{1}{2!}\sum_{j=1}^{n}\sum_{i=1}^{n}\frac{\partial G(x)}{\partial x_i}\Bigg|_{x=x^*}(x_i - x_i^*)(x_j - x_j^*) + \text{H.O.T.} \tag{AII.2}$$

By ignoring the second-order term and the higher-order terms (H.O.T.), and because the MPP is on the failure surface so that $G(x)|_{x=x^*} = 0$, equation (2.11) becomes

$$G(x) \approx \sum_{i=1}^{n}\frac{\partial G(x)}{\partial x_i}\Bigg|_{x=x^*}(x_i - x_i^*). \tag{AII.3}$$

The simplified limit state function can then be transformed into a standard space by using the standard normal variables, denoted as $z = (z_1, \ldots, z_n)$, where $z_i = (x_i - \mu_i)/\sigma_i$. Substituting into $x_i - x_i^*$ gives

$$x_i - x_i^* = (\sigma_i z_i - \mu_i) - (\sigma_i z_i^* - \mu_i) = \sigma_i(z_i - z_i^*) \qquad \text{(AII.4)}$$

and

$$\frac{\partial G(x)}{\partial x_i} = \frac{\partial G(z)}{\partial z_i}\left(\frac{\partial z_i}{\partial x_i}\right) = \frac{1}{\sigma_i}\frac{\partial G(z)}{\partial z_i}. \qquad \text{(AII.5)}$$

Substituting equations (AII.4) and (AII.5) into (AII.3) yields

$$G(z) \approx \sum_{i=1}^{n} \left.\frac{\partial G(z)}{\partial z_i}\right|_{z=z^*} (z_i - z_i^*). \qquad \text{(AII.6)}$$

As $\mu_{z_i} = 0, \sigma_{z_i} = 1$, the mean and variance of the limit state function $G(z)$ can be found:

$$\mu_G \approx -\sum_{i=1}^{n} z_i^* \left.\frac{\partial G(z)}{\partial z_i}\right|_{z=z^*} \qquad \text{(AII.7)}$$

$$\sigma_G^2 \approx \sum_{i=1}^{n} \left(\frac{\partial G(z)}{\partial z_i}\right)^2\bigg|_{z=z^*}. \qquad \text{(AII.8)}$$

The safety index is then obtained as

$$\beta_{HL} = \frac{\mu_G}{\sigma_G} = \frac{-\sum_{i=1}^{n} z_i^* \left.\frac{\partial G(z)}{\partial z_i}\right|_{z=z^*}}{\sqrt{\sum_{i=1}^{n}\left(\frac{\partial G(z)}{\partial z_i}\right)^2\bigg|_{z=z^*}}}, \qquad \text{(AII.9)}$$

which is equation (AII.1).

The design point is given by

$$z_i^* = \alpha_i^* \beta_{HL} \quad (i = 1, 2, \ldots, n), \qquad \text{(AII.10)}$$

where

$$\alpha_i^* = \frac{\left.\frac{\partial G(z)}{\partial z_i}\right|_{z=z^*}}{\sqrt{\sum_{i=1}^{n}\left(\frac{\partial G(z)}{\partial z_i}\right)^2\bigg|_{z=z^*}}} \qquad \text{(AII.11)}$$

are the direction cosines along the coordinate axis z.

In the space of the original variables, the design point is

$$x_i = \mu_i - \alpha_i^* \sigma_i \beta_{HL}. \qquad \text{(AII.12)}$$

The above techniques can be used to perform iteration processes to compute the HL safety index and the design point until convergence is obtained.

Modal Analysis of Dynamic Systems [77,78]

A typical multiple DOF dynamic system (total number of DOFs is n), excited by external forces, can be described by spatial property matrices as

$$[M]\{\ddot{x}(t)\} + [C]\{\dot{x}(t)\} + [K]\{x(t)\} = \{f(t)\}, \tag{AIII.1}$$

where $[M]$, $[C]$, and $[K]$ are $n \times n$ symmetric mass, viscous damping, and stiffness matrices respectively. $\{\ddot{x}(t)\}$, $\{\dot{x}(t)\}$, $\{x(t)\}$, and $\{f(t)\}$ are $n \times 1$ vectors of time-varying acceleration, velocity, displacement responses, and external excitation forces respectively.

The unique property of the system, its own natural mode of vibration, can be obtained under the consideration of the undamped and free vibration of the system:

$$[M]\{\ddot{x}(t)\} + [K]\{x(t)\} = 0. \tag{AIII.2}$$

The solutions of $\{x(t)\}$ are of the form:

$$\{x(t)\} = \{\overline{X}\}e^{i\omega t}, \tag{AIII.3}$$

where $\{\overline{X}\}$ is the $n \times 1$ vector of time-independent response amplitudes, and the velocity and acceleration are

$$\{\dot{x}(t)\} = i\omega\{\overline{X}\}e^{i\omega t} \tag{AIII.4}$$

$$\{\ddot{x}(t)\} = -\omega^2\{\overline{X}\}e^{i\omega t}. \tag{AIII.5}$$

Substituting the above equations into (AIII.2) gives

$$[[K] - \omega^2[M]]\{\overline{X}\} = 0. \tag{AIII.6}$$

In order for equation (AIII.6) to have a nontrivial solution, the inverse of $[[K]-\omega^2[M]]$ must not exist. Therefore, the determinant of $[[K]-\omega^2[M]]$ should satisfy

$$\det[[K] - \omega^2[M]] = 0. \tag{AIII.7}$$

Equation (AIII.7) is called the *characteristic equation*. The eigen solutions are the paramount characteristics of the dynamic system. They are (1) n positive real eigenvalues, $\omega_1^2, \omega_2^2, \ldots, \omega_n^2$, of which the square root values

169

ω_1, ω_2, ..., ω_n are undamped *natural frequencies*, and (2) n corresponding $n \times 1$ eigenvectors $\{\psi_1\}$, $\{\psi_2\}$, ..., $\{\psi_n\}$, which are the *mode shapes*. The complete solutions are

$$[\omega_r^2] = \begin{bmatrix} \omega_1^2 & 0 & \cdots & 0 \\ 0 & \omega_2^2 & \cdots & 0 \\ \cdots & \cdots & \cdots & \cdots \\ 0 & 0 & \cdots & \omega_n^2 \end{bmatrix} \tag{AIII.8}$$

and

$$\{\psi\} = [\{\psi_1\}, \quad \{\psi_2\}, \ldots, \quad \{\psi_n\}]. \tag{AIII.9}$$

Each pair ω_r^2 and $\{\psi_r\}$ is called a *vibration mode* of the system.

The dynamic system also has special and important properties called *orthogonality properties*, which are described as

$$\begin{aligned} [\Phi]^T[M][\Phi] &= [I] \\ [\Phi]^T[K][\Phi] &= [\omega_r^2], \end{aligned} \tag{AIII.10}$$

or in terms of a single element,

$$\begin{aligned} \phi_j{}^T[M]\phi_j &= 1 \\ \phi_j{}^T[K]\phi_j &= \omega_j^2, \end{aligned} \tag{AIII.11}$$

where $[\Phi]$ is the *mass-normalized modal matrix* and ϕ_j is its jth column vector, in detail

$$\phi_j = \begin{pmatrix} \phi_{j,1} \\ \phi_{j,2} \\ \cdots \\ \phi_{j,s} \\ \cdots \\ \phi_{j,n} \end{pmatrix}_{n \times 1}, \tag{AIII.12}$$

and

$$\phi_j^T = (\phi_{j,1} \quad \phi_{j,2} \quad \cdots \quad \phi_{j,r} \quad \cdots \quad \phi_{j,n})_{1 \times n}. \tag{AIII.13}$$

Further, taking the excitation force vector $\{f(t)\} = \{\overline{F}\}e^{i\omega t}$, where $\{\overline{F}\}$ is the force amplitude constant, and substituting it, together with equations (AIII.4) and (AIII.5), into equation (AIII.1), gives

$$([K] - \omega^2[M] + i\omega[C])\{\overline{X}\}e^{i\omega t} = \{\overline{F}\}e^{i\omega t}. \tag{AIII.14}$$

Solving for $\{\overline{X}\}$ yields

$$\{\overline{X}\} = ([K] - \omega^2[M] + i\omega[C])^{-1}\{\overline{F}\} = [\alpha(\omega)]\{\overline{F}\}, \tag{AIII.15}$$

where $[\alpha(\omega)]$ is the *receptance FRF (frequency response function) matrix*, which contains all the information of the system dynamic characteristics.

Quite often, the response model is presented in terms of modal coordinates. For a multiple DOF dynamic system with n total DOFs, if the dissipative mechanism is hysteretic[1] and damping is proportional, in the case of forced harmonic vibration, the equation of motion can be expressed as

$$[M]\{\ddot{x}(t)\} + i[D]\{x(t)\} + [K]\{x(t)\} = \{f(t)\}, \tag{AIII.16}$$

where $[D]$ is the $n \times n$ hysteretic damping matrix.

Any response vector such as $\{x\}$ can be expressed as a linear combination of the eigenvectors,[2] i.e. the corresponding mass-normalized mode shapes,

$$\{x(t)\} = \{\overline{X}\}e^{i\omega t} = \sum_{r=1}^{n} a_r \phi_r = [\Phi]\{a\} \tag{AIII.17}$$

$$\{\ddot{x}(t)\} = -\omega^2 \{\overline{X}\}e^{i\omega t} = \sum_{r=1}^{n} \ddot{a}_r \phi_r = [\Phi]\{\ddot{a}\}. \tag{AIII.18}$$

Substituting equations (AIII.17) and (AIII.18) into equation (AIII.16) gives

$$[M][\Phi]\{\ddot{a}\} + i[D][\Phi]\{a\} + [K][\Phi]\{a\} = \{f(t)\}. \tag{AIII.19}$$

Pre-multiplying by $[\Phi]^T$ on both sides of equation (AIII.19) yields

$$[\Phi]^T[M][\Phi]\{\ddot{a}\} + i[\Phi]^T[D][\Phi]\{a\} + \Phi^T[K][\Phi]\{a\} = [\Phi]^T\{f(t)\}. \tag{AIII.20}$$

Applying the orthogonality properties to equation (AIII.20) yields

$$[I]\{\ddot{a}\} + i[\eta\omega^2]\{a\} + [\omega^2]\{a\} = [\Phi]^T\{f(t)\}. \tag{AIII.21}$$

Extracting one element from the above equation results in the following description:

$$\ddot{a}_r + i\eta_r \omega_r^2 a_r + \omega_r^2 a_r = \phi_r^T \{f(t)\}. \tag{AIII.22}$$

Assuming that the solution of the above equation is of the form $a_r = \overline{a}_r e^{i\omega t}$ and $\{f(t)\} = \{\overline{F}\}e^{i\omega t}$, substituting them into equation (AIII.22) yields

---

1. A hysteretic damping model is adopted hereafter in this book, because of the advantage of describing more closely the energy dissipation mechanism exhibited by most real structures in cases of forced vibrations, as well as of providing a much simpler analysis for multiple DOF systems [77].

2. Due to a linearly independent set of vectors in n space possessing orthogonality properties.

$$-\omega^2 \bar{a}_r + i\eta_r \omega_r^2 \bar{a}_r + \omega_r^2 \bar{a}_r = \phi_r^T \{\overline{F}\}. \tag{AIII.23}$$

Solving for $\bar{a}_r$ gives

$$\bar{a}_r = \frac{\phi_r^T \{\overline{F}\}}{(\omega_r^2 - \omega^2) + i\eta_r \omega_r^2}. \tag{AIII.24}$$

Substituting equation (AIII.24) into equation (AIII.17) leads to

$$\{x(t)\} = \{\overline{X}\}e^{i\omega t} = \sum_{r=1}^{n} \frac{\{\phi\}_r^T \{\overline{F}\}\{\phi\}_r}{(\omega_r^2 - \omega^2) + i\eta_r \omega_r^2} e^{i\omega t} \tag{AIII.25}$$

and finally to

$$\{\overline{X}\} = \sum_{r=1}^{n} \frac{\{\phi\}_r^T \{\overline{F}\}\{\phi\}_r}{(\omega_r^2 - \omega^2) + i\eta_r \omega_r^2}. \tag{AIII.26}$$

Given that all the spatial property elements are known variables, the displacement response vector can be solved through a finite element program. The other dynamic characteristics, e.g. the natural frequencies and mode shapes, can be obtained through eigen solutions.

To perform the reliability analysis of the system, at a required location for example, the safety margin can then be constructed by the response/characteristics information obtained through the above modeling techniques, followed by relevant reliability analysis methods.

Multiple Force Analysis

If m harmonic forces are introduced, denoted as F_{k_1}, F_{k_2}, ..., F_{k_m}, sharing the same excitation frequency ω and applied to nodes k_1, k_2, ..., k_m respectively, equation (3.2) becomes

$$\overline{X}_j = \sum_{r=1}^{n} \frac{\phi_{r,j} \sum_{k=k_1}^{k_m} F_k \phi_{r,k}}{(\omega_r^2 - \omega^2) + i\eta_r \omega_r^2}. \tag{AIV.1}$$

Applying the same simplification theory as was used in Section 3.1 yields (AIV.2), which is similar to equation (3.13):

$$|\overline{X}_j|^2 \approx \sum_{r=1}^{n} \frac{\left| \phi_{r,j} \sum_{k=k_1}^{k_m} F_k \phi_{r,k} \right|^2}{(\omega_r^2 - \omega^2)^2 + (\eta_r \omega_r^2)^2}. \tag{AIV.2}$$

Again, the complex *numerator* of (AIV.1) can be defined as $g_{j,r}$, i.e.

$$g_{j,r} = \phi_{r,j} \sum_{k=k_1}^{k_m} F_k \phi_{r,k}. \tag{AIV.3}$$

Assuming that changes are made to the defined parameter $g_{j,r}$ with respect to small changes of spatial parameters, equation (AIV.3) becomes

$$g_{j,r} + \Delta g_{j,r} = (\phi_{r,j} + \Delta \phi_{r,j}) \sum_{k=k_1}^{k_m} F_k (\phi_{r,k} + \Delta \phi_{r,k}). \tag{AIV.4}$$

Expanding the above equation and ignoring the higher (second)-order terms yields

$$\Delta g_{j,r} = \Delta \phi_{r,j} \sum_{k=k_1}^{k_m} F_k \phi_{r,k} + \phi_{r,j} \sum_{k=k_1}^{k_m} F_k \Delta \phi_{r,k}. \tag{AIV.5}$$

Equation (AIV.5) represents a generic case of one element in the vector of the defined parameter $\Delta g_{j,r}$ with respect to j and r, which is denoted as G_j, i.e.

$$G_j = \Delta g_j = \begin{pmatrix} \Delta g_{j,1} \\ \Delta g_{j,2} \\ \cdots \\ \Delta g_{j,n} \end{pmatrix}. \tag{AIV.6}$$

G_j can be deduced as

$$G_j = \begin{pmatrix} \Delta\phi_{1,j} & 0 & \cdots & 0 \\ 0 & \Delta\phi_{2,j} & \cdots & \cdots \\ \cdots & \cdots & \cdots & \cdots \\ \cdots & \cdots & \cdots & \Delta\phi_{n,j} \end{pmatrix}_{n\times n} \begin{pmatrix} \phi_{1,k_1} & \phi_{1,k_2} & \cdots & \phi_{1,k_m} \\ \phi_{2,k_2} & \phi_{2,k_2} & \cdots & \cdots \\ \cdots & \cdots & \cdots & \cdots \\ \phi_{n,k_m} & \phi_{n,k_m} & \cdots & \phi_{n,k_m} \end{pmatrix}_{n\times m} \begin{pmatrix} F_{k_1} \\ F_{k_2} \\ \cdots \\ F_{k_m} \end{pmatrix}_{m\times 1}$$

$$+ \begin{pmatrix} \phi_{1,j} & 0 & \cdots & 0 \\ 0 & \phi_{2,j} & \cdots & \cdots \\ \cdots & \cdots & \cdots & \cdots \\ \cdots & \cdots & \cdots & \phi_{n,j} \end{pmatrix}_{n\times n} \begin{pmatrix} \Delta\phi_{1,k_1} & \Delta\phi_{1,k_2} & \cdots & \Delta\phi_{1,k_m} \\ \Delta\phi_{2,k_2} & \Delta\phi_{2,k_2} & \cdots & \cdots \\ \cdots & \cdots & \cdots & \cdots \\ \Delta\phi_{n,k_m} & \Delta\phi_{n,k_m} & \cdots & \Delta\phi_{n,k_m} \end{pmatrix}_{n\times m} \begin{pmatrix} F_{k_1} \\ F_{k_2} \\ \cdots \\ F_{k_m} \end{pmatrix}_{m\times 1}. \tag{AIV.7}$$

In the above equation, the term

$$\begin{pmatrix} \Delta\phi_{1,j} & 0 & \cdots & 0 \\ 0 & \Delta\phi_{2,j} & \cdots & \cdots \\ \cdots & \cdots & \cdots & \cdots \\ \cdots & \cdots & \cdots & \Delta\phi_{n,j} \end{pmatrix}_{n\times n}$$

is the diagonal matrix of the defined vector $\tilde{\Phi}_j$, and

$$\tilde{\Phi}_j = \Delta\tilde{\phi}_j = \begin{pmatrix} \Delta\phi_{1,j} \\ \Delta\phi_{2,j} \\ \cdots \\ \Delta\phi_{n,j} \end{pmatrix}_{n\times 1}, \tag{AIV.8}$$

which was given by equation (3.82).

Another term,

$$\begin{pmatrix} \Delta\phi_{1,k_1} & \Delta\phi_{1,k_2} & \cdots & \Delta\phi_{1,k_m} \\ \Delta\phi_{2,k_2} & \Delta\phi_{2,k_2} & \cdots & \cdots \\ \cdots & \cdots & \cdots & \cdots \\ \Delta\phi_{n,k_m} & \Delta\phi_{n,k_m} & \cdots & \Delta\phi_{n,k_m} \end{pmatrix}_{n\times m},$$

is the matrix consisting of m columns of $\tilde{\Phi}_k$ ($k = k_1, \ldots, k_m$), i.e.

$$
\begin{pmatrix}
\Delta\phi_{1,k_1} & \Delta\phi_{1,k_2} & \cdots & \Delta\phi_{1,k_m} \\
\Delta\phi_{2,k_2} & \Delta\phi_{2,k_2} & \cdots & \cdots \\
\cdots & \cdots & \cdots & \cdots \\
\Delta\phi_{n,k_m} & \Delta\phi_{n,k_m} & \cdots & \Delta\phi_{n,k_m}
\end{pmatrix}_{n \times m}
= (\tilde{\Phi}_{k_1} \quad \tilde{\Phi}_{k_2} \quad \cdots \quad \tilde{\Phi}_{k_m}). \quad \text{(AIV.9)}
$$

Thus, according to equation (3.80) and the definitions from (3.82)–(3.90), the following definitions are made:

$$
\tilde{S} =
\begin{pmatrix}
\phi_{1,k_1} & \phi_{1,k_2} & \cdots & \phi_{1,k_m} \\
\phi_{2,k_2} & \phi_{2,k_2} & \cdots & \cdots \\
\cdots & \cdots & \cdots & \cdots \\
\phi_{n,k_m} & \phi_{n,k_m} & \cdots & \phi_{n,k_m}
\end{pmatrix}_{n \times m}
\quad \text{(AIV.10)}
$$

$$
F =
\begin{pmatrix}
F_{k_1} \\
F_{k_2} \\
\cdots \\
F_{k_m}
\end{pmatrix}_{m \times 1}
\quad \text{(AIV.11)}
$$

$$
J =
\begin{pmatrix}
\Delta\phi_{1,j} & 0 & \cdots & 0 \\
0 & \Delta\phi_{2,j} & \cdots & \cdots \\
\cdots & \cdots & \cdots & \cdots \\
\cdots & \cdots & \cdots & \Delta\phi_{n,j}
\end{pmatrix}_{n \times n}
= \sum_{i=1}^{n} t_i \tilde{\Phi}_j Z_i, \quad \text{(AIV.12)}
$$

where $t_i = \begin{bmatrix} 0 & \cdots & 1_i & \cdots & 0 \end{bmatrix}_{1 \times n}$ is a vector with only the ith element being 1 (all others are zero),

$$
Z_i =
\begin{bmatrix}
0 & \cdots & \cdots & \cdots & 0 \\
\cdots & \cdots & \cdots & \cdots & \cdots \\
\cdots & \cdots & 1_i & \cdots & \cdots \\
\cdots & \cdots & \cdots & \cdots & \cdots \\
0 & \cdots & \cdots & \cdots & 0
\end{bmatrix}_{n \times n}
$$

is a diagonal matrix with only the ith element being 1 (all others are zero), and

$$
Y_k =
\begin{pmatrix}
\Delta\phi_{1,k_1} & \Delta\phi_{1,k_2} & \cdots & \Delta\phi_{1,k_m} \\
\Delta\phi_{2,k_2} & \Delta\phi_{2,k_2} & \cdots & \cdots \\
\cdots & \cdots & \cdots & \cdots \\
\Delta\phi_{n,k_m} & \Delta\phi_{n,k_m} & \cdots & \Delta\phi_{n,k_m}
\end{pmatrix}
= \sum_{i=1}^{m} \tilde{\Phi}_{k_i} t_i. \quad \text{(AIV.13)}
$$

Therefore, equation (AIV.7) can be rewritten in matrix form:

$$
G_j = S_j Y_k F + J \tilde{S} F = S_j \left(\sum_{i}^{m} \tilde{\Phi}_{k_i} t_i \right) F + \left(\sum_{i}^{n} t_i \tilde{\Phi}_j Z_i \right) \tilde{S} F, \quad \text{(AIV.14)}
$$

where S_j is defined in equation (3.80).

Considering equation (3.89), which is

$$\tilde{\Phi}_j = \tilde{Q}_j k - \tilde{P}_j m - \tilde{W}_j m, \tag{AIV.15}$$

and substituting it into equation (AIV.14) gives

$$G_j = S_j \left(\sum_i^m (\tilde{Q}_{k_i} k - \tilde{P}_{k_i} m - \tilde{W}_{k_i} m) t_i \right) F + \left(\sum_i^n t_i (\tilde{Q}_j k - \tilde{P}_j m - \tilde{W}_j m) Z_i \right) \tilde{S} F. \tag{AIV.16}$$

Because F is complex, G_j is complex and can be rewritten separately in real and imaginary parts as

$$G_j = \begin{pmatrix} G_j^R \\ G_j^I \end{pmatrix}, \tag{AIV.17}$$

where

$$G_j^R = S_j \left(\sum_i^m (\tilde{Q}_{k_i} k - \tilde{P}_{k_i} m - \tilde{W}_{k_i} m) t_i \right) F^R + \left(\sum_i^n t_i (\tilde{Q}_j k - \tilde{P}_j m - \tilde{W}_j m) Z_i \right) \tilde{S} F^R \tag{AIV.18}$$

$$G_j^I = S_j \left(\sum_i^m (\tilde{Q}_{k_i} k - \tilde{P}_{k_i} m - \tilde{W}_{k_i} m) t_i \right) F^I + \left(\sum_i^n t_i (\tilde{Q}_j k - \tilde{P}_j m - \tilde{W}_j m) Z_i \right) \tilde{S} F^I. \tag{AIV.19}$$

Therefore, the covariance matrix of G_j can be derived as

$$C_{G_j} = E[G_j (G_j)^T] = E\left[\begin{pmatrix} G_j^R \\ G_j^I \end{pmatrix} \left((G_j^R)^T \quad (G_j^I)^T \right) \right] = \begin{pmatrix} C_{G_j^R (G_j^R)^T} & C_{G_j^R (G_j^I)^T} \\ C_{G_j^I (G_j^R)^T} & C_{G_j^I (G_j^I)^T} \end{pmatrix}, \tag{AIV.20}$$

where

$$C_{G_j^R (G_j^R)^T} = E[G_j^R (G_j^R)^T] =$$
$$E\left[\begin{array}{l} \left(S_j \left(\sum_i^m (\tilde{Q}_{k_i} k - \tilde{P}_{k_i} m - \tilde{W}_{k_i} m) t_i \right) F^R + \left(\sum_i^n t_i (\tilde{Q}_j k - \tilde{P}_j m - \tilde{W}_j m) Z_i \right) \tilde{S} F^R \right) \bullet \\ \left(S_j \left(\sum_i^m (\tilde{Q}_{k_i} k - \tilde{P}_{k_i} m - \tilde{W}_{k_i} m) t_i \right) F^R + \left(\sum_i^n t_i (\tilde{Q}_j k - \tilde{P}_j m - \tilde{W}_j m) Z_i \right) \tilde{S} F^R \right)^T \end{array} \right] \tag{AIV.21}$$

$$C_{G_j^I(G_j^I)^T} = E[G_j^I(G_j^I)^T] =$$

$$E\left[\begin{array}{l}\left(S_j\left(\sum_i^m(\tilde{Q}_{k_i}k - \tilde{P}_{k_i}m - \tilde{W}_{k_i}m)t_i\right)F^I + \left(\sum_i^n t_i(\tilde{Q}_jk - \tilde{P}_jm - \tilde{W}_jm)Z_i\right)\tilde{S}_kF^I\right)\bullet \\ \left(S_j\left(\sum_i^m(\tilde{Q}_{k_i}k - \tilde{P}_{k_i}m - \tilde{W}_{k_i}m)t_i\right)F^I + \left(\sum_i^n t_i(\tilde{Q}_jk - \tilde{P}_jm - \tilde{W}_jm)Z_i\right)\tilde{S}_kF^I\right)^T\end{array}\right]$$

$$(\text{AIV}.22)$$

and

$$C_{G_j^R(G_j^I)^T} = E[G_j^R(G_j^I)^T] =$$

$$E\left[\begin{array}{l}\left(S_j\left(\sum_i^m(\tilde{Q}_{k_i}k - \tilde{P}_{k_i}m - \tilde{W}_{k_i}m)t_i\right)F^R + \left(\sum_i^n t_i(\tilde{Q}_jk - \tilde{P}_jm - \tilde{W}_jm)Z_i\right)\tilde{S}F^R\right)\bullet \\ \left(S_j\left(\sum_i^m(\tilde{Q}_{k_i}k - \tilde{P}_{k_i}m - \tilde{W}_{k_i}m)t_i\right)F^I + \left(\sum_i^n t_i(\tilde{Q}_jk - \tilde{P}_jm - \tilde{W}_jm)Z_i\right)\tilde{S}F^I\right)^T\end{array}\right]$$

$$(\text{AIV}.23)$$

Therefore, the covariance of G_j can be evaluated if C_k and C_m are known. The expected value of the defined parameter $g_{j,r}$ can be derived as

$$E[g_{j,r}] = E\left[\phi_{r,j}\left(\sum_{k=k_1}^{k_m} F_k\phi_{r,k}\right)\right] = E\left[\sum_{k=k_1}^{k_m} F_k\phi_{r,k}\phi_{r,j}\right]. \qquad (\text{AIV}.24)$$

Again, it can be expressed in two parts:

$$E[g_{j,r}^R] = E\left[\sum_{k=k_1}^{k_m} F_k^R\phi_{r,k}\phi_{r,j}\right] = \sum_{k=k_1}^{k_m} F_k^R(E[\phi_{r,k}]E[\phi_{r,j}] + Cov(\phi_{r,k}, \phi_{r,j}))$$

$$(\text{AIV}.25)$$

$$E[g_{j,r}^I] = E\left[\sum_{k=k_1}^{k_m} F_k^I\phi_{r,k}\phi_{r,j}\right] = \sum_{k=k_1}^{k_m} F_k^I(E[\phi_{r,k}]E[\phi_{r,j}] + Cov(\phi_{r,k}, \phi_{r,j})).$$

$$(\text{AIV}.26)$$

So far, all the required statistical information, when multiple forces are involved, can be obtained for reliability analysis.

Summary of the Defined Parameters

Modal Parameters

Parameters	ω_j^2	ϕ_j
Parameter alternative definition	$\Delta\omega_j^2 = \phi_j^T \Delta K \phi_j - \phi_j^T \Delta M \phi_j \omega_j^2$ $\Omega = (\Delta\omega_1^2 \quad \Delta\omega_2^2 \quad \cdots \quad \Delta\omega_n^2)^T$	$\Delta\phi_j = \sum_{k=1,k\neq j}^{n} a_k\phi_k + a_j\phi_j = \sum_{k\neq j}\left[\dfrac{\phi_k[-\omega_j^2\Delta M + \Delta K]\phi_j}{\omega_j^2 - \omega_k^2}\right]\phi_k - \tfrac{1}{2}(\phi_j^T\Delta M\phi_j)\phi_j$ $\Phi_j = (\Delta\phi_{j,1} \quad \Delta\phi_{j,2} \quad \cdots \quad \Delta\phi_{j,n})^T$
Other definition	$k = (\Delta k_{11} \quad \Delta k_{21} \quad \cdots \quad \Delta k_{n1} \quad \cdots \quad \Delta k_{nn})^T$ $m = (\Delta m_{11} \quad \Delta m_{21} \quad \cdots \quad \Delta m_{n1} \quad \cdots \quad \Delta m_{nn})^T$ $\phi_j = (\phi_{j,1} \quad \phi_{j,2}, \quad \cdots \quad \phi_{j,s} \quad \cdots \quad \phi_{j,n})^T$ $A = \begin{pmatrix} \phi_{1,1}\phi_{1,1} & \phi_{1,2}\phi_{1,1} & \cdots & \phi_{1,0}\phi_{1,1} & \cdots & \phi_{1,n}\phi_{1,n} \\ \phi_{2,1}\phi_{2,1} & \phi_{2,2}\phi_{2,1} & & & & \phi_{2,n}\phi_{2,n} \\ \vdots & \vdots & & \phi_{j,s}\phi_{j,s} & & \vdots \\ \vdots & \vdots & & & & \vdots \\ \phi_{n,1}\phi_{n,1} & \phi_{n,2}\phi_{n,1} & \cdots & \phi_{n,0}\phi_{n,1} & \cdots & \phi_{n,n}\phi_{n,n} \end{pmatrix}$	$\alpha_{jk} = \dfrac{\omega_j^2}{\omega_j^2 - \omega_k^2}$ $\beta_{jk} = \dfrac{1}{\omega_j^2 - \omega_k^2}$ $P_j = \begin{pmatrix} \sum_{k\neq j}^{n}\alpha_{jk}\phi_{k,1}\phi_{j,1}\phi_{k,1} & \sum_{k\neq j}^{n}\alpha_{jk}\phi_{k,2}\phi_{j,1}\phi_{k,1} & \cdots & \sum_{k\neq j}^{n}\alpha_{jk}\phi_{k,n}\phi_{j,1}\phi_{k,1} & \cdots & \sum_{k\neq j}^{n}\alpha_{jk}\phi_{k,n}\phi_{j,n}\phi_{k,1} \\ \sum_{k\neq j}^{n}\alpha_{jk}\phi_{k,1}\phi_{j,1}\phi_{k,2} & & & \sum_{k\neq j}^{n}\alpha_{jk}\phi_{k,n}\phi_{j,s}\phi_{k,1} & & \\ \vdots & & & \vdots & & \vdots \\ \sum_{k\neq j}^{n}\alpha_{jk}\phi_{k,1}\phi_{j,1}\phi_{k,n} & \cdots & & & \cdots & \sum_{k\neq j}^{n}\alpha_{jk}\phi_{k,n}\phi_{j,n}\phi_{k,n} \end{pmatrix}$

$$B = \begin{pmatrix} \omega_1^2\phi_{1,1}\phi_{1,1} & \omega_1^2\phi_{1,2}\phi_{1,1} & \cdots & \omega_1^2\phi_{1,n}\phi_{1,1} \\ \omega_2^2\phi_{2,1}\phi_{2,1} & \omega_2^2\phi_{2,2}\phi_{2,2} & \cdots & \omega_2^2\phi_{2,n}\phi_{2,n} \\ \vdots & \vdots & \omega_j^2\phi_{j,r}\phi_{j,s} & \vdots \\ \vdots & \vdots & \cdots & \vdots \\ \omega_n^2\phi_{n,1}\phi_{n,1} & \omega_n^2\phi_{n,2}\phi_{n,1} & \omega_n^2\phi_{n,n}\phi_{n,1} & \omega_n^2\phi_{n,n}\phi_{n,n} \end{pmatrix}$$

$$Q_j = \begin{pmatrix} \sum_{k\neq j}\beta_{jk}\phi_{k,1}\phi_{j,1}\phi_{k,1} & \sum_{k\neq j}^n\beta_{jk}\phi_{k,2}\phi_{j,1}\phi_{k,1} & \cdots & \sum_{k\neq j}^n\beta_{jk}\phi_{k,r}\phi_{j,1}\phi_{k,1} & \cdots & \sum_{k\neq j}^n\beta_{jk}\phi_{k,n}\phi_{j,1}\phi_{k,1} \\ \sum_{k\neq j}\beta_{jk}\phi_{k,1}\phi_{j,1}\phi_{k,2} & & & & & \vdots \\ \vdots & & & & & \vdots \\ \sum_{k\neq j}^n\beta_{jk}\phi_{k,1}\phi_{j,1}\phi_{k,n} & & & \sum_{k\neq j}^n\beta_{jk}\phi_{k,r}\phi_{j,s}\phi_{k,l} & & \sum_{k\neq j}^n\beta_{jk}\phi_{k,n}\phi_{j,1}\phi_{k,n} \end{pmatrix}$$

$$W_j = \frac{1}{2}\begin{pmatrix} \phi_{j,1}\phi_{j,1}\phi_{j,1} & \phi_{j,2}\phi_{j,1}\phi_{j,1} & \cdots & \phi_{j,n}\phi_{j,1}\phi_{j,1} \\ \phi_{j,1}\phi_{j,1}\phi_{j,2} & \phi_{j,2}\phi_{j,1}\phi_{j,2} & \cdots & \phi_{j,n}\phi_{j,1}\phi_{j,2} \\ \vdots & \vdots & \cdots & \vdots \\ \phi_{j,1}\phi_{j,1}\phi_{j,n} & & \cdots & \phi_{j,n}\phi_{j,1}\phi_{j,n} \end{pmatrix}_{N\times N^2}$$

Matrix form

$$\Delta\omega_j^2 = -\omega_j^2 \sum_s^n\sum_r^n [\phi_{j,r}\phi_{j,s}]\Delta m_{sr} + \sum_s^n\sum_r^n [\phi_{j,r}\phi_{j,s}]\Delta k_{rs}$$

$$\Omega = Ak - Bm$$

$$\Delta\phi_{j,r} = \sum_r^n\sum_s^n\left[\sum_{k\neq j}^n -\alpha_{jk}\phi_{k,s}\phi_{j,r}\phi_{k,r}\right]\Delta m_{sr} + \sum_r^n\sum_s^n\left[\sum_{k\neq j}\beta_{jk}\phi_{k,s}\phi_{j,r}\phi_{k,r}\right]\Delta k_{sr}$$

$$\Phi_j = Q_j k - P_j m - W_j m$$

Covariance matrix

$$C_\Omega = AC_k A^T + BC_m B^T$$

$$C_{\theta_j} = Q_j C_k Q_j^T + P_j C_m P_j^T + P_j C_m W_j^T + W_j C_m P_j^T + W_j C_m W_j^T$$

Defined Parameters

Parameter definition	$d_r = (\omega_r^2 - \omega^2)^2 + (\eta_r \omega_r^2)^2$	$r_{jk,r} = \phi_{r,j} F_k \phi_{r,k}$

Alternative definition	$\Delta d_r = 2(\omega_r^2 - \omega^2 + \eta_r^2 \omega_r^2)\Delta\omega_r^2$	$\Delta r_{jk,r} = \phi_{r,j} F_k \Delta\phi_{r,k} + \Delta\phi_{r,j} F_k \phi_{r,k}$
	$D = (\Delta d_1 \quad \Delta d_2 \quad \cdots \quad \Delta d_n)^T$	$R_{jk} = (\Delta r_{jk,1} \quad \Delta r_{jk,2} \quad \cdots \quad \Delta r_{jk,n})^T$

Other definition

$$H = 2\begin{pmatrix} \omega_1^2 - \omega^2 + \eta_1^2\omega_1^2 & 0 & \cdots & 0 \\ 0 & \omega_2^2 - \omega^2 + \eta_2^2\omega_2^2 & \cdots & \vdots \\ \vdots & \vdots & \cdots & \vdots \\ \vdots & \vdots & \cdots & \omega_n^2 - \omega^2 + \eta_n^2\omega_n^2 \end{pmatrix}$$

$$S_j = \begin{pmatrix} \phi_{1,j} & 0 & \cdots & 0 \\ 0 & \phi_{2,j} & \cdots & \vdots \\ \vdots & \vdots & \cdots & \vdots \\ \vdots & \vdots & \cdots & \phi_{n,j} \end{pmatrix}_{N\times N}$$

$$S_k = \begin{pmatrix} \phi_{1,k} & 0 & \cdots & 0 \\ 0 & \phi_{2,k} & \cdots & \vdots \\ \vdots & \vdots & \cdots & \vdots \\ \vdots & \vdots & \cdots & \phi_{n,k} \end{pmatrix}_{N\times N}$$

$$\bar{P}_j = \begin{pmatrix} \sum_{k\neq1}^n \alpha_{1k}\phi_{k,1}\phi_{1,1}\phi_{k,j} & \sum_{k\neq1}^n \alpha_{1k}\phi_{k,2}\phi_{1,1}\phi_{k,j} & \cdots & \sum_{k\neq1}^n \alpha_{1k}\phi_{k,1}\phi_{1,s}\phi_{k,j} & \cdots & \sum_{k\neq1}^n \alpha_{1k}\phi_{k,n}\phi_{1,n}\phi_{k,j} \\ \sum_{k\neq2}^n \alpha_{2k}\phi_{k,1}\phi_{2,1}\phi_{k,j} & \sum_{k\neq2}^n \alpha_{2k}\phi_{k,2}\phi_{2,1}\phi_{k,j} & \cdots & & \cdots & \sum_{k\neq2}^n \alpha_{2k}\phi_{k,n}\phi_{2,n}\phi_{k,j} \\ \vdots & \vdots & & \sum_{k\neq j}^n \alpha_{jk}\phi_{k,s}\phi_{j,s}\phi_{k,j} & & \vdots \\ \sum_{k\neq n}^n \alpha_{nk}\phi_{k,1}\phi_{n,1}\phi_{k,j} & \sum_{k\neq n}^n \alpha_{nk}\phi_{k,2}\phi_{n,1}\phi_{k,j} & \cdots & & \cdots & \sum_{k\neq n}^n \alpha_{nk}\phi_{k,n}\phi_{n,n}\phi_{k,j} \end{pmatrix}_{N\times N^2}$$

$$\bar{Q}_l = \begin{pmatrix} \sum_{\substack{k\neq 1}}^{n}\beta_{1k}\phi_{k,1}\phi_{1,1}\phi_{k,j} & \sum_{\substack{k\neq 1}}^{n}\beta_{1k}\phi_{k,2}\phi_{1,1}\phi_{k,j} & \cdots & \sum_{\substack{k\neq 1}}^{n}\beta_{1k}\phi_{k,r}\phi_{1,s}\phi_{k,l} & \cdots & \sum_{\substack{k\neq 1}}^{n}\beta_{1k}\phi_{k,n}\phi_{1,n}\phi_{k,j} \\ \sum_{\substack{k\neq 2}}^{n}\beta_{2k}\phi_{k,1}\phi_{1,1}\phi_{k,j} & \sum_{\substack{k\neq 2}}^{n}\beta_{2k}\phi_{k,1}\phi_{2,1}\phi_{k,j} & \cdots & \cdots & \cdots & \sum_{\substack{k\neq 2}}^{n}\beta_{2k}\phi_{k,n}\phi_{2,n}\phi_{k,j} \\ \cdots & \cdots & \cdots & \sum_{\substack{k\neq j}}^{n}\beta_{jk}\phi_{k,r}\phi_{j,s}\phi_{k,l} & \cdots & \cdots \\ \cdots & \cdots & \cdots & \cdots & \cdots & \cdots \\ \sum_{\substack{k\neq n}}^{n}\beta_{nk}\phi_{k,1}\phi_{n,1}\phi_{k,j} & \sum_{\substack{k\neq n}}^{n}\beta_{nk}\phi_{k,1}\phi_{n,1}\phi_{k,j} & \cdots & \cdots & \cdots & \sum_{\substack{k\neq n}}^{n}\beta_{nk}\phi_{k,n}\phi_{n,n}\phi_{k,j} \end{pmatrix}_{N\times N^2}$$

$$\bar{W}_j = \frac{1}{2}\begin{pmatrix} \phi_{1,1}\phi_{1,1}\phi_{1,j} & \phi_{1,2}\phi_{1,1}\phi_{1,j} & \cdots & \phi_{1,n}\phi_{1,n}\phi_{1,j} \\ \phi_{2,1}\phi_{2,1}\phi_{2,j} & \phi_{2,2}\phi_{2,1}\phi_{2,j} & \cdots & \phi_{2,n}\phi_{2,n}\phi_{2,j} \\ \cdots & \cdots & \cdots & \cdots \\ \cdots & \cdots & \cdots & \cdots \\ \phi_{n,1}\phi_{n,1}\phi_{n,j} & \phi_{n,2}\phi_{n,1}\phi_{n,j} & \cdots & \phi_{n,n}\phi_{n,n}\phi_{n,j} \end{pmatrix}_{N\times N^2}$$

Matrix form $D = H\Omega$

$$R_{jk} = r_k(S_j\bar{\Phi}_k + S_k\bar{\Phi}_j) = r_k[S_j(\bar{Q}_k - \bar{P}_k m - \bar{W}_k m) + S_k(\bar{Q}_j k - \bar{P}_j m - \bar{W}_j m)]$$

Covariance matrix

$$C_D = H(AC_kA^T + BC_mB^T)H$$

$$C_{R_k} = r_k^2\Big[S_j\big(C_{\bar{\Phi}_j}S_j + S_k E\big(\bar{\Phi}_k\bar{\Phi}_k^T\big)S_j + S_j E\big(\bar{\Phi}_k\bar{\Phi}_k^T\big)S_k + S_k C_{\bar{\Phi}_j}S_k\big]$$

$$= r_k^2\left[\begin{aligned} & S_k\big(\bar{Q}_k C_k\bar{Q}_k^T + \bar{P}_k C_m\bar{P}_k^T + \bar{P}_k C_m\bar{W}_k^T + \bar{W}_k C_m\bar{P}_k^T + \bar{W}_k C_m\bar{W}_k^T\big)S_j \\ & + S_k\big(\bar{Q}_k C_k\bar{Q}_k^T + \bar{P}_j C_m\bar{P}_k^T + \bar{P}_k C_m\bar{P}_k^T + \bar{W}_j C_m\bar{P}_k^T + \bar{W}_j C_m\bar{W}_k^T\big)S_j \\ & + S_j\big(\bar{Q}_k C_k\bar{Q}_j^T + \bar{P}_k C_m\bar{W}_j^T + \bar{P}_k C_m\bar{P}_j^T + \bar{W}_k C_m\bar{P}_j^T + \bar{W}_k C_m\bar{W}_j^T\big)S_k \\ & + S_k\big(\bar{Q}_k C_k\bar{Q}_j^T + \bar{P}_j C_m\bar{P}_j^T + \bar{P}_j C_m\bar{P}_j^T + \bar{W}_j C_m\bar{P}_j^T + \bar{W}_j C_m\bar{W}_j^T\big)S_k \end{aligned}\right]$$

Combined Parameters

Parameter definition	$T = (D \quad R_{jk})^T = (\Delta d_1 \quad \Delta d_2 \quad \cdots \quad \Delta d_n \quad \Delta r_{jk,1} \quad \Delta r_{jk,2} \quad \cdots \quad \Delta r_{jk,n})^T$
Other definition	$T_{jk} = \left(\underset{HA}{F_k(S_j \tilde{Q}_k + S_k \tilde{Q}_j)} \right)_{2N \times N^2} k - \left(\underset{HB}{F_k(S_j \tilde{P}_k + S_j \tilde{W}_k + S_k \tilde{P}_j + S_k \tilde{W}_j)} \right)_{2N \times N^2} m$
	$U_{jk} = \left(\underset{HA}{F_k(S_j \tilde{Q}_k + S_k \tilde{Q}_j)} \right)_{2N \times N^2}$
	$V_{jk} = \left(\underset{HB}{F_k(S_j \tilde{P}_k + S_j \tilde{W}_k + S_k \tilde{P}_j + S_k \tilde{W}_j)} \right)_{2N \times N^2}$
Matrix form	$T_{jk} = U_{jk} k - V_{jk} m$
Covariance matrix	$C_{T_{jk}} = U_{jk} C_k U_{jk}^T + V_{jk} C_m V_{jk}^T$

Nodal Coordinates of the Helicopter Model

Node Number	X-Coordinate (m)	Y-Coordinate (m)	Z-Coordinate (m)
1	−1.50	−0.60	−0.25
2	−1.50	−0.60	0.25
3	−1.50	0.60	0.25
4	−1.50	0.60	−0.25
5	0.00	−1.00	−0.80
6	0.00	−1.00	0.80
7	0.00	1.00	−0.80
8	0.00	1.00	0.80
9	0.50	−1.00	0.00
10	0.50	0.00	−0.80
11	0.50	0.00	0.80
12	0.50	1.00	0.00
13	1.00	−1.00	−0.80
14	1.00	−1.00	0.80
15	1.00	0.00	0.80
16	1.00	1.00	−0.80
17	1.00	1.00	0.80
18	1.50	−1.00	0.00
19	1.50	0.00	−0.80
20	1.50	0.00	0.80
21	1.50	1.00	0.00
22	2.00	−1.00	−0.80
23	2.00	−1.00	0.80
24	2.00	1.00	−0.80
25	2.00	1.00	0.80
26	4.50	0.00	0.50
27	6.00	−1.00	0.50
28	6.00	0.00	0.50
29	6.00	1.00	0.50
30	6.50	0.00	0.50
31	6.80	0.00	1.50

Element Connectivity and Properties of the Helicopter Model

Elem. No.	Connectivity		A (m^2)	$Iyy = Izz$ (m^4)	J (m^4)	Height_y (m)	Height_z (m)
	From Node	To Node					
1	1	2	0.224×10^{-3}	2.942×10^{-8}	5.884×10^{-8}	0.03	0.03
2	1	3	0.224×10^{-3}	2.942×10^{-8}	5.884×10^{-8}	0.03	0.03
3	1	5	0.224×10^{-3}	2.942×10^{-8}	5.884×10^{-8}	0.03	0.03
4	2	4	0.224×10^{-3}	2.942×10^{-8}	5.884×10^{-8}	0.03	0.03
5	2	6	0.224×10^{-3}	2.942×10^{-8}	5.884×10^{-8}	0.03	0.03
6	3	4	0.224×10^{-3}	2.942×10^{-8}	5.884×10^{-8}	0.03	0.03
7	3	7	0.224×10^{-3}	2.942×10^{-8}	5.884×10^{-8}	0.03	0.03
8	4	8	0.224×10^{-3}	2.942×10^{-8}	5.884×10^{-8}	0.03	0.03
9	5	6	0.224×10^{-3}	2.942×10^{-8}	5.884×10^{-8}	0.03	0.03
10	5	7	0.224×10^{-3}	2.942×10^{-8}	5.884×10^{-8}	0.03	0.03
11	5	9	0.224×10^{-3}	2.942×10^{-8}	5.884×10^{-8}	0.03	0.03
12	5	10	0.224×10^{-3}	2.942×10^{-8}	5.884×10^{-8}	0.03	0.03
13	5	13	0.224×10^{-3}	2.942×10^{-8}	5.884×10^{-8}	0.03	0.03
14	6	8	0.224×10^{-3}	2.942×10^{-8}	5.884×10^{-8}	0.03	0.03
15	6	9	0.224×10^{-3}	2.942×10^{-8}	5.884×10^{-8}	0.03	0.03
16	6	11	0.224×10^{-3}	2.942×10^{-8}	5.884×10^{-8}	0.03	0.03
17	6	14	0.224×10^{-3}	2.942×10^{-8}	5.884×10^{-8}	0.03	0.03
18	7	8	0.224×10^{-3}	2.942×10^{-8}	5.884×10^{-8}	0.03	0.03
19	7	10	0.224×10^{-3}	2.942×10^{-8}	5.884×10^{-8}	0.03	0.03

Elem. No.	Connectivity		A (m^2)	$Iyy = Izz$ (m^4)	J (m^4)	Height_y (m)	Height_z (m)
	From Node	To Node					
20	7	12	0.224×10^{-3}	2.942×10^{-8}	5.884×10^{-8}	0.03	0.03
21	7	16	0.224×10^{-3}	2.942×10^{-8}	5.884×10^{-8}	0.03	0.03
22	8	11	0.224×10^{-3}	2.942×10^{-8}	5.884×10^{-8}	0.03	0.03
23	8	12	0.224×10^{-3}	2.942×10^{-8}	5.884×10^{-8}	0.03	0.03
24	8	17	0.224×10^{-3}	2.942×10^{-8}	5.884×10^{-8}	0.03	0.03
25	9	13	0.224×10^{-3}	2.942×10^{-8}	5.884×10^{-8}	0.03	0.03
26	9	14	0.224×10^{-3}	2.942×10^{-8}	5.884×10^{-8}	0.03	0.03
27	10	13	0.224×10^{-3}	2.942×10^{-8}	5.884×10^{-8}	0.03	0.03
28	6	10	0.224×10^{-3}	2.942×10^{-8}	5.884×10^{-8}	0.03	0.03
29	11	14	0.224×10^{-3}	2.942×10^{-8}	5.884×10^{-8}	0.03	0.03
30	11	17	0.224×10^{-3}	2.942×10^{-8}	5.884×10^{-8}	0.03	0.03
31	12	16	0.224×10^{-3}	2.942×10^{-8}	5.884×10^{-8}	0.03	0.03
32	12	17	0.224×10^{-3}	2.942×10^{-8}	5.884×10^{-8}	0.03	0.03
33	13	14	0.900×10^{-3}	3.075×10^{-7}	6.150×10^{-7}	0.05	0.05
34	13	16	0.900×10^{-3}	3.075×10^{-7}	6.150×10^{-7}	0.05	0.05
35	13	18	0.224×10^{-3}	2.942×10^{-8}	5.884×10^{-7}	0.03	0.03
36	13	19	0.224×10^{-3}	2.942×10^{-8}	5.884×10^{-7}	0.03	0.03
37	13	22	0.224×10^{-3}	2.942×10^{-8}	5.884×10^{-7}	0.03	0.03
38	14	15	0.900×10^{-3}	3.075×10^{-7}	6.150×10^{-7}	0.05	0.05
39	14	18	0.224×10^{-3}	2.942×10^{-8}	5.884×10^{-7}	0.03	0.03
40	14	20	0.224×10^{-3}	2.942×10^{-8}	5.884×10^{-7}	0.03	0.03
41	14	23	0.224×10^{-3}	2.942×10^{-8}	5.884×10^{-7}	0.03	0.03
42	15	17	0.900×10^{-3}	3.075×10^{-7}	6.150×10^{-7}	0.05	0.05
43	16	17	0.900×10^{-3}	3.075×10^{-7}	6.150×10^{-7}	0.05	0.05
44	16	19	0.224×10^{-3}	2.942×10^{-8}	5.884×10^{-7}	0.03	0.03
45	16	21	0.224×10^{-3}	2.942×10^{-8}	5.884×10^{-7}	0.03	0.03
46	16	24	0.224×10^{-3}	2.942×10^{-8}	5.884×10^{-7}	0.03	0.03
47	17	20	0.224×10^{-3}	2.942×10^{-8}	5.884×10^{-7}	0.03	0.03
48	17	21	0.224×10^{-3}	2.942×10^{-8}	5.884×10^{-7}	0.03	0.03
49	17	25	0.224×10^{-3}	2.942×10^{-8}	5.884×10^{-7}	0.03	0.03

Elem. No.	Connectivity		A (m^2)	$Iyy = Izz$ (m^4)	J (m^4)	Height_y (m)	Height_z (m)
	From Node	To Node					
50	18	22	0.224×10^{-3}	2.942×10^{-8}	5.884×10^{-7}	0.03	0.03
51	18	23	0.224×10^{-3}	2.942×10^{-8}	5.884×10^{-7}	0.03	0.03
52	19	22	0.224×10^{-3}	2.942×10^{-8}	5.884×10^{-7}	0.03	0.03
53	19	24	0.224×10^{-3}	2.942×10^{-8}	5.884×10^{-7}	0.03	0.03
54	20	23	0.224×10^{-3}	2.942×10^{-8}	5.884×10^{-7}	0.03	0.03
55	20	25	0.224×10^{-3}	2.942×10^{-8}	5.884×10^{-7}	0.03	0.03
56	21	24	0.224×10^{-3}	2.942×10^{-8}	5.884×10^{-7}	0.03	0.03
57	21	25	0.224×10^{-3}	2.942×10^{-8}	5.884×10^{-7}	0.03	0.03
58	22	23	0.224×10^{-3}	2.942×10^{-8}	5.884×10^{-7}	0.03	0.03
59	22	24	0.224×10^{-3}	2.942×10^{-8}	5.884×10^{-7}	0.03	0.03
60	22	26	0.224×10^{-3}	2.942×10^{-8}	5.884×10^{-7}	0.03	0.03
61	23	25	0.224×10^{-3}	2.942×10^{-8}	5.884×10^{-7}	0.03	0.03
62	23	26	0.224×10^{-3}	2.942×10^{-8}	5.884×10^{-7}	0.03	0.03
63	24	25	0.224×10^{-3}	2.942×10^{-8}	5.884×10^{-7}	0.03	0.03
64	24	26	0.224×10^{-3}	2.942×10^{-8}	5.884×10^{-7}	0.03	0.03
65	25	26	0.224×10^{-3}	2.942×10^{-8}	5.884×10^{-7}	0.03	0.03
66	26	28	1.492×10^{-3}	1.688×10^{-6}	3.376×10^{-6}	0.10	0.10
67	27	28	1.492×10^{-3}	1.688×10^{-6}	3.376×10^{-6}	0.10	0.10
68	28	29	1.492×10^{-3}	1.688×10^{-6}	3.376×10^{-6}	0.10	0.10
69	28	30	1.492×10^{-3}	1.688×10^{-6}	3.376×10^{-6}	0.10	0.10
70	30	31	1.492×10^{-3}	1.688×10^{-6}	3.376×10^{-6}	0.10	0.10

E (N/m^2) $= 7.0 \times 10^{10}$, Rho (kg/m^3) $= 2800$, $Nu = 0.35$, for all elements.

References

[1] J. Heyman, Structural analysis, A Historical Approach, Cambridge University Press, Cambridge, UK, 1998.

[2] D. Rowland, T.N. Howe, Vitruvius, Ten Books on Architecture, Cambridge University Press, Cambridge, UK, 1999.

[3] B. Hanson, Architects and the 'Building World' from Chambers to Ruskin, Cambridge University Press, Cambridge, UK, 2003.

[4] L. Lao, Material Mechanics in Ancient China, University Press, University of Science and Technology for National Defence, Changsha, P.R. China, 1991.

[5] H.O. Madsen, S. Krenk, N.C. Lind, Methods of Structural Safety, Prentice-Hall, 1990.

[6] I. Elishakeoff, Probabilistic Methods in the Theory of Structures, Wiley, New York, 1983.

[7] R.E. Melchers, Structural Reliability Analysis and Prediction, second ed., Wiley, 1999.

[8] C.S. Manohar, S. Gupta, Modelling and evaluation of structural reliability: current status and future directions, in: K.S. Jagadish, R.N. Iyenagar (Eds.), Recent Advances in Structural Engineering, University Press, Hyderabad, India, 2005, pp. 90–187.

[9] R.S. Langley, Unified approach to probabilistic and possibilistic analysis of uncertain systems, J. Eng. Mech. 126 (11) (2000) 1163–1172.

[10] S. Ulam, Adventures of a Mathematician, Charles Scribner's Sons, New York, 1983.

[11] G.S. Fishman, Monte Carlo, Concepts, Algorithms, and Applications, Springer-Verlag, New York, 1996.

[12] C.P. Robert, G. Casell, Monte Carlo Statistical Methods, second ed., Springer-Verlag, New York, 2004.

[13] J. Schneider, Introduction to Safety and Reliability of Structures, International Association for Bridge and Structural Engineering (IABSE), Switzerland, 1997.

[14] W.Q. Zhu, Y.J. Ren, W.Q. Wu, A stochastic FEM based on local averages of random vector fields, J. Eng. Mech. 118 (3) (1992) 496–511.

[15] G. Maymon, Direct computation of the design point of a stochastic structure using a finite element code, Struct. Saf. 14 (1994) 185–202.

[16] C.E. Brenner, C. Bucher, A contribution to the SFE-based reliability assessment of nonlinear structures under dynamic loading, J. Eng. Mech. 13 (1) (1998) 265–273.

[17] B.M. Ayyub, C.Y. Chia, Generalized conditional expectation for structural reliability assessment, Struct. Saf. 11 (1992) 131–146.

[18] A. Haldar, S. Mahadevan, First-order/second-order reliability methods (FORM/SORM), in: C. Sundararajan (Ed.), Probabilistic Structural Mechanics Handbook, Chapman & Hall, New York, 1993, pp. 27–52. Chapter 3

[19] D.G. Robinson, Sandia Report: A Survey of Probabilistic Methods Used in Reliability, Risk and Uncertainty Analysis: Analytical Techniques I, Sandia National Laboratories, USA, 1998.

[20] A.C. Cornell, A probability based structural code, ACI J. Proc. 66 (12) (1969).

[21] O. Ditlevsen, Structural reliability and the invariance problem, Report No. 22, Solid Mechanics Division, University of Waterloo, Canada, 1973.

[22] A.M. Hasofer, N.C. Lind, Exact and invariant second-moment code format, Proc. ASCE, J. Eng. Mech. Div., ASCE 100 (EM1) (1974).

[23] R. Rackwitz, B. Fiessler, Structural reliability under combined random load sequences, Comput. Struct. 9 (1978) 489–494.

[24] M. Hohenbichler, R. Rackwitz, Non-normal dependent vectors in structural safety, J. Eng. Mech. Div., ASCE 107 (EM6) (1981) 1227–1238.

[25] D. Moens, D. Vandepitte, An overview of novel non-probabilistic approaches for non-deterministic dynamic analysis, Keynote lecture on NOise and Vibration: Emerging Methods, NOVEM 2005, Saint-Raphal, paper 158, 2005.

[26] R. Moore, Interval Analysis, Prentice Hall, Englewood Cliffs, 1966.

[27] G. Alefeld, J. Herzberger, Introduction to Interval Computation, Academic Press, New York, 1983.

[28] A. Neumaier, Interval Methods for Systems of Equations, Cambridge University Press, Cambridge, UK, 1990.

[29] E. Hansen, Global Optimisation using Interval Analysis, Marcel Dekker, New York, 1992.

[30] I. Elishakoff, Some Questions in Eigenvalue Problems in Engineering, International Series of Numerical Mathematics, vol. 96, Birkhauser-Verlag, Basel, Switzerland, 1991, pp. 71–107

[31] H.U. Koyluoglu, A.S. Cakmak, R.K. Nielsen, Interval algebra to deal with pattern loading and structural uncertainties, J. Eng. Mech. 121 (11) (1995).

[32] O. Dessombz, F. Thouverez, J.-P. Laine, L. Jezequel, Analysis of mechanical systems using interval computations applied to finite element method, J. Sound Vib. 239 (5) (2001) 949–968.

[33] S.S. Rao, L. Berke, Analysis of uncertain structural systems using interval analysis, AIAA J. 35 (4) (1997).

[34] L. Zadeh, Fuzzy sets, Inf. Control 8 (1965) 338–353.

[35] L. Zadeh, Fuzzy sets as a basis for a theory of possibility, Fuzzy Sets Syst. 1 (1978) 3–28.

[36] D. Driankov, H. Hellendoorn, M. Reinfrank, An Introduction to Fuzzy Control, Springer-Verlag, Berlin, 1996.

[37] R.L. Mullen, R.L. Muhanna, Bounds of structural response for all possible loading combinations, J. Struct. Eng. 125 (1) (1999).

[38] Z. Lei, Q. Chen, A new approach to fuzzy finite element analysis, J. Comput. Methods Appl. Mech. Eng. 191 (2002) 5113–5118.

[39] L. Chen, S.S. Rao, Fuzzy finite-element approach for the vibration analysis of imprecisely-defined systems, J. Finite Elem. Anal. Des. 27 (1997) 69–83.

[40] T.M. Wasfy, A.K. Noor, Application of fuzzy sets to transist analysis of space structures, J. Finite Elem. Anal. Des. 29 (1998) 153–171.

[41] A. Cherki, G. Plessis, B. Lallemand, T. Tison, P. Level, Fuzzy behaviour of mechanical systems with uncertain boundary conditions, J. Comput. Methods Appl. Mech. Eng. 189 (2000) 863–873.

[42] S.S. Rao, L. Cao, Fuzzy boundary element method for the analysis of imprecisely defined systems, AIAA J. 39 (9) (2001).

[43] G. Venter, R.T. Haftka, Using response surface approximations in fuzzy set based design optimisation, in: Proc. 39th AIAA/ASME/ASCE/AHS/ASC Structures, Structural Dynamics, Materials Conference, Long Beach, CA, April 20–23, 1998.

[44] R.H. Myers, D.C. Montgomery, Response Surface Methodology – Process and Product Optimisation Using Designed Experiments, second ed., Wiley, 2002.

[45] R.H. Myers, Response surface methodology — Current status and future directions, J. Qual. Technol. 31 (1) (1999) 30–44.

[46] G.E.P. Box, N.R. Draper, Empirical Model-Building and Response Surface, Wiley, 1987.

[47] A.I. Khuri, J.A. Cornell, revised and expanded Response Surface: Design and Analyses, second ed., Marcel Dekker, 1996

[48] R.H. Myers, A.I. Khuri, W.H.C. Carter, Response surface methodology: 1966–1988, Technometrics 31 (2) (1989) 137–157.

[49] W.J. Hill, W.G. Hunter, A review of response surface methodology: a literature survey, Technometrics 8 (4) (1996) 571–590.

[50] C.G. Bucher, U. Bourgund, A fast and efficient response surface approach for structural reliability problems, Struct. Saf. 7 (1990) 57–66.

[51] M.R. Rajashekhar, B.R. Ellingwood, A new look at the response surface approach for reliability analysis, Struct. Saf. 12 (1993) 205–220.

[52] M.R. Rajashekhar, B.R. Ellingwood, Reliability of reinforced-concrete cylindrical shells, J. Struct. Eng. 121 (2) (1995) 336–347.

[53] Y.W. Liu, F. Moses, A sequential response surface method and its application in the reliability analysis of aircraft structural systems, Struct. Saf. 16 (1994) 39–46.

[54] S.H. Kim, S.W. Na, Response surface method using vector projected sampling, Struct. Saf. 19 (1) (1997) 3–19.

[55] Y.G. Zhao, T. Ono, System reliability evaluation of ductile frame structures, J. Struct. Eng. 124 (6) (1998) 678–685.

[56] J. Huh, P.E.A. Haldar, Stochastic finite-element-based seismic risk of nonlinear structures, J. Struct. Eng. 127 (3) (2001) 323–329.

[57] R.C. Soares, A. Mohamed, W.S. Venturini, M. Lemaire, Reliability analysis of non-linear reinforced concrete frames using response surface method, Reliab. Eng. Syst. Saf. 75 (2002) 1–16.

[58] V.J. Remero, L.P. Swiler, A.A. Giunta, Application of finite element, global polynomial, and kriging response surface in progressive lattice sampling designs, in: Proc. 8th ASCE Specialty Conference on Probabilistic Mechanics and Structural Reliability, PMC2000-175, 2000.

[59] X.L. Guan, R.E. Melcher, A parametric study on the response surface method, in: Proc. 8th ASCE Specialty Conference on Probabilistic Mechanics and Structural Reliability, PMC2000-023, 2000.

[60] X.L. Guan, R.E. Melcher, Effect of response surface parameter variation on structural reliability estimates, Struct. Saf. 23 (2001) 429–444.

[61] T.W. Simpson, T.M. Mauery, J. Korte, F. Mistree, Kriging models for global approximation in simulation-based multidisciplinary design optimization, AIAA J. 39 (12) (2001) 2233–2241.

[62] J.J.M. Rijpkema, A.J.G. Schoofs, L.F.P. Etman, RSM based design optimization, J. Math. Mech. 81 (6) (2001) 691–692.

[63] G.G. Wang, Z.M. Dong, P. Aitchison, Adaptive response surface method — A global optimization scheme for approximation-based design problems, Eng. Optim. 33 (2001) 707–733.

[64] T.T. Allen, L.Y. Yu, Low-cost response surface methods from simulation optimisation, Qual. Reliab. Eng. Int. 8 (2002) 5–17.

[65] L. Faravelli, Response-surface approach for reliability analysis, J. Eng. Mech. 115 (12) (1989) 2763–2781.

[66] K. Breitung, L. Faravelli, Response surface methods and asymptotic approximations, in: F. Casciati, et al. (Eds.), Mathematical Models for Structural Reliability Analysis, Part 5, CRC Press, 1996.

[67] F. Bohm, A. Bruckner-Foit, On criteria for accepting a response surface model, Probab. Eng. Mech. 7 (1992) 183−190.

[68] T.H-J. Yao, Y.-K. Wen, Response surface method for time-variant reliability analysis, J. Struct. Eng. 122 (2) (1996) 193−201.

[69] F.S. Wong, Uncertainties in dynamic soil-structure interaction, J. Eng. Mech. 110 (2) (1984) 308−325.

[70] F.S. Wong, Slope reliability and response surface method, J. Geotech. Eng. 111 (1) (1985) 32−53.

[71] A.M. Brown, A.A.F. Ferri, Application of the probabilistic dynamic synthesis method to realistic structures, AIAA J. 37 (10) (1999) 1292−1297.

[72] Y.G. Zhao, T. Ono, H. Idota, Response uncertainty and time-variant reliability analysis for hysteretic and MDF structures, Earthquake Eng. Struct. Dyn. 28 (1999) 1187−1213.

[73] B. Wu, An Investigation of Response Surface Methodology (RSM) for Structural Reliability Analysis, First Year Report, Engineering Department, University of Cambridge, 2003.

[74] X. Qu, R. Haftka, Probabilistic safety factor based response surface approach for reliability-based design optimisation, Third ISSMO/AIAA Internet Conference on Approximations in Optimisation, October 14−25, 2002.

[75] N. Gayton, J.M. Bourinet, M. Lemaire, CQ2RS: a new statistical approach to the response surface method for reliability analysis, J. Struct. Saf. 25 (2003) 99−121.

[76] B.D. Youn, K.K. Choi, A new response surface methodology for reliability-based design optimisation, J. Comput. Struct. 82 (2004) 241−256.

[77] N.M.M. Maia, J.M.M. Silva, Theoretical and Experimental Modal Analysis, Research Studies Press, Taunton, UK, 1997.

[78] D.J. Ewins, Modal Testing, Theory, Practice and Application, second ed., Research Studies Press, Taunton, UK, 2000.

[79] R.S. Langley, P.J. Shorter, V. Cotoni, Predicting the Response Statistics of Uncertain Structures Using Extended Versions of SEA, 2005 Congress and Exposition on Noise Control Engineering, Rio de Janeiro, Brazil, August 7−10, 2005.

[80] R. Langley, P. Bremner, A hybrid-method for the vibration analysis of complex structural acoustic systems, J. Acoust. Soc. Am. 105 (3) (1999) 1657−1671.

[81] V. Cotoni, R.S. Langley, M.R.F. Kidner, Numerical and experimental validation of variance prediction in the statistical energy analysis of built-up systems, J. Sound Vib. 288 (2005) 701−728.

[82] A.D. Dimarogonas, Interval analysis of vibrating systems, J. Sound Vib. 183 (4) (1995) 739−749.

[83] Z. Qiu, S.H. Chen, I. Elishakoff, Bounds of eigenvalues for structures with an interval description of uncertain-but-non-random parameters, Chaos, Solutions Fractals 7 (1996) 425−434.

[84] Z. Qiu, S.H. Chen, I. Elishakeoff, Non-probabilistic eigenvalue problem for structures with uncertain parameters via interval analysis, Chaos, Solutions and Fractals 7 (1996) 303−308.

[85] Z. Qiu, I. Elishakoff, J.H. Starnes Jr, The bound set of possible eigenvalues of structures with uncertain but non-random parameters, Chaos, Solutions Fractals 11 (1996) 1845−1857.

[86] A. Deif, Advanced Matrix Theory for Scientists and Engineers, second ed., Abacus Press, Tunbridge Wells, UK, 1991.

[87] S.H. Chen, X.M. Zhang, Y.D. Chen, Interval eigenvalues of closed-loop systems of uncertain structures, Comput. Struct. 84 (2006) 243–253.

[88] S.H. Chen, J. Wu, Interval optimization of dynamic response for structures with interval parameters, Comput. Struct. 82 (2004) 1–11.

[89] T.M. Wasfy, A.K. Noor, Finite element analysis of flexible multibody systems with fuzzy parameters, J. Comput. Methods Appl. Mech. Eng. 160 (1998) 223–243.

[90] B. Lallemand, A. Cherki, T. Tison, P. Level, Fuzzy modal finite element analysis of structures with imprecise material properties, J. Sound Vib. 220 (2) (1999) 353–364.

[91] D. Moens, D. Vandepitte, Fuzzy finite element method for frequency response function analysis of uncertain structures, AIAA J. 40 (1) (2002).

[92] D. Moens, D. Vandepitte, A survey of non-probabilistic uncertainty treatment in finite element analysis, J. Comput. Methods Appl. Mech. Eng. 194 (2005) 1527–1555.

[93] M.D. Munck, D. Moens, W. Desmet, D. Vandepitte, An automated procedure for interval and fuzzy finite element analysis, in: Proc. ISMA 2004, International Conference on Noise and Vibration Engineering, Leuven, Belgium, 2004, pp. 3023–3033.

[94] D. Moens, D. Vandepitte, Envelope frequency response function calculation of uncertain structures, in: Proc. 25th International Conference on Noise and Vibration Engineering, ISMA 25, vol. I, Leuven, Belgium, 2000, pp. 395–402.

[95] D. Moens, D. Vandepitte, W. Teichert, Non-probabilistic approaches for non-deterministic dynamic FE analysis of imprecisely defined structures, in: Proc. ISMA 2004, International Conference on Noise and Vibration Engineering,, Leuven, Belgium, 2004, pp. 3095–3119.

[96] D. Moens, D. Vandepitte, W. Teichert, Application of the fuzzy finite element method in structural dynamics, in: Proc. 23rd International Conference on Noise and Vibration Engineering, ISMA 23, vol. II, Leuven, Belgium, 1998, pp. 975–982.

[97] H.D. Gersem, D. Moens, W. Desmet, D. Vandepitte, Interval and fuzzy finite element analysis of mechanical structures with uncertain parameters, in: Proc. ISMA 2004, International Conference on Noise and Vibration Engineering, Leuven, Belgium, 2004, pp. 3009–3021.

[98] S. Donders, D. Vandepitte, J. Van de Peer, W. Desmet, The short transformation method to predict the FRF of dynamic structures subject to uncertainty, in: Proc. ISMA 2004, International Conference on Noise and Vibration Engineering, Leuven, Belgium, 2004, pp. 3043–3054.

[99] M.F. Pellissetti, G.I. Schueller, H.J. Pradlwarter, A. Calvi, S. Fransen, M. Klein, Reliability analysis of spacecraft structures under static and dynamic loading, Comput. Struct. 84 (2006) 1313–1325.

[100] J. Margetson, Technical Report: Vibration Analysis of Aircraft Structures Using a Response Surface Representation of the Dynamic Finite Element Analysis, Defence Research Consultancy, UK, August 2001.

[101] J. Margetson, Technical Report: Application of Probabilistic Response Surface Procedures to the Vibration Analysis of Aircraft Structures, Defence Research Consultancy, UK, January 2001.

[102] P.B. Nair, A.J. Keane, Stochastic reduced basis methods, AIAA J. 40 (8) (2002) 1653–1664.

[103] E. Nikolaidis, Chapter 17 – Probabilistic analysis of dynamic systems, in: E. Nikolaidis, D.M. Ghiocel, S. Singhal, N. Nikolaidis (Eds.), Engineering Design Reliability Handbook, CRC Press, 2004.

[104] B. Wu, A Combined Approach to Reliability Analysis of Dynamic Systems, Second Year Report, Engineering Department, University of Cambridge, 2004.

[105] R.Y. Rubinstein, Simulation and the Monte-Carlo Method, Wiley, New York, 1981.

[106] J.S. Bendat, A.G. Piersol, Random Data: Analysis and Measurement Procedures, third ed., Wiley, 2000.

[107] M.L. Shooman, Probabilistic Reliability: An Engineering Approach, McGraw-Hill, New York, 1968.

[108] M. Shinozuka, Basis analysis of structural safety, J. Struct. Div., ASCE 3 (109) (1983).

[109] A. Ang, W.H. Tang, Probability Concepts in Engineering Planning and Design – Volume II: Decision, Risk and Reliability, Wiley, New York, 1984.

[110] G.I. Schueller, R.S. Stix, A critical appraisal of methods to determine failure probabilities, Struct. Saf. 4 (1987) 293–309.

[111] Y.G. Zhao, T. Ono, New approximations for SORM: part 1, J. Eng. Mech. 125 (1) (1999) 79–85.

[112] Y.G. Zhao, T. Ono, New approximations for SORM: part 2, J. Eng. Mech. 125 (1) (1999) 86–93.

[113] Y.G. Zhao, T. Ono, A general procedure for first/second-order reliability method (FORM/SORM), Struct. Saf. 21 (1999) 95–112.

[114] Y.G. Zhao, T. Ono, Moment methods for structural reliability, Struct. Saf. 3 (2001) 47–75.

[115] F.F. Yap, J. Woodhouse, Investigation of damping dffects on statistical energy analysis of coupled structures, J. Sound Vib. 197 (3) (1996) 351–371.

[116] A. Mathews, V.R. Sule, C. Venkatesan, Order reduction and closed-loop vibration control in helicopter fuselage, J. Guide. Control, Dyn. 25 (2) (2002).

[117] S. Newman, The Foundation of Helicopter Flight, Edward Arnold, London, UK, 1994.

[118] R. Ganguli, Survey of recent developments in rotorcraft design optimization, J. Aircr. 41 (3) (2004) 493–510.

[119] M. Yang, I. Chopra, D. Haas, Vibration prediction for rotor system with faults using coupled rotor-fuselage model, J. Aircr. 41 (2) (2004) 348–358.

[120] L. Monterrubio, I. Sharf, Influence of landing-gear design on helicopter ground resonance, Can. Aeronaut. Space J. (CASJ) 48 (2) (2002).

[121] C. Venkatesan, A. Udayasankar, Selection of sensor locations for active vibration control of helicopter fuselage, J. Aircr. 36 (2) (1999) 434–442.

[122] J.H. McMasters, R.M. Cummings, From farther, faster, higher to leaner, meaner, greener: further directions in aeronautics, J. Aircr. 41 (1) (2004) 51–61.

[123] D.R. Ballal, J. Zelina, Progress in aeroengine technology (1939–2003), J. Aircr. 41 (1) (2004) 43–50.

[124] D.E. Heverly II, K.W. Wang, E.C. Smith, An optimal actuator placement methodology for active control of helicopter airframe vibrations, J. Am. Helicopter Soc. (2001) 251–261.

[125] P.P. Friedmann, A. Millott, Vibration reduction in rotorcraft using active control: A comparison of various approaches, Journal of Guidance, Control, Dyn. 18 (4) (1995).

[126] M.D. Rao, Recent applications of viscoelastic damping for noise control in automobiles and commercial airplanes, J. Sound Vib. 262 (3) (2003) 457–474.

[127] D.P. Richards, M.P. Neale, Effects of phase relationship in helicopter vibrations, J.IES (1994) 26–31.

[128] T.S. Murthy, Optimisation of helicopter airframe structures for vibration reduction – consideration, formulations, and applications, J. Aircr. 28 (1) (1991) 66–73.

[129] S.P. King, The minimisation of helicopter vibration through blade design and active control, Aeronautical J. (1988) 247–263.

[130] W. Berczynski, W.B. Woodson, Fail safe evaluation of helicopter structures, in: Proc. 6th Joint FAA/DoD/NASA Aging Aircraft Conference, September 16–19, 2002.

[131] W.J. Twomey, T.L.C. Chen, I.U. Ojalvo, T. Ting, A general method for modifying a finite-element model to correlate with modal test data, J. Am. Helicopter Soc. (1991) 48–58.

[132] M. Revivo, O. Rand, Feasibility of non-rotating active lifting surface for helicopter vibration reduction, J. Smart Mater. Struct. 10 (2001) 154–171.

[133] R. Ganguli, Optimum design of a helicopter rotor for low vibration using aeroelastic analysis and response surface methods, J. Sound Vib. 258 (2) (2002) 327–344.

[134] S. Mangallck, C. Venkatesan, N.N. Kishore, Technical Report: IITK/ARDB/AVCH/826/ 95/01, Department of Aerospace Engineering, India Institute of Technology, Kanpur, India, May 1995.

[135] M. Urban, Analysis of the fatigue life of riveted sheet metal helicopter airframe joints, Int. J. Fatigue 25 (2003) 1013–1026.

[136] R.G. Kvaternik, The NASA/industry design analysis methods for vibrations (DAMVIBS) program – Accomplishments and contributions, presented at American Helicopter Society National Technical Specialists' Meeting on Rotorcraft Structures, Williamsburg, Virginia, October 29–31, 1991.

[137] H. Guan, R. Gibson, Micromechanical models for damping in woven fabric-reinforced polymer matrix composites, J. Compos. Mater. 35 (16) (2001) 1417–1434.

[138] B. Panda, E. Mychalowycz, F.J. Tarzanin, Application of passive dampers to modern helicopters, J. Smart Mater. Struct. 5 (1996) 509–516.

[139] R.D. Cook, Finite Element Modelling for Stress Analysis, Wiley, 1995.

Index

Note: Page numbers followed by "*f*" and "*t*" refer to figures and tables, respectively.

Printed and bound by CPI Group (UK) Ltd, Croydon, CR0 4YY

03/10/2024

01040422-0001